W9-AUN-662
3 1611 00312 2410

AN INTRODUCTION TO STATISTICAL MECHANICS

An Introduction to Statistical Mechanics

By

PHILIPPE DENNERY

Laboratoire de Physique Théorique et des Hautes Energies
Faculté des Sciences, Orsay

A Halsted Press Book

JOHN WILEY & SONS
New York

FIRST PUBLISHED IN 1972

Published in the U.S.A. by
Halsted Press, a Division of
John Wiley & Sons, Inc.
New York.

ISBN 0 470-20910-0

Library of Congress Catalog Card Number: 72-4150

Printed in Great Britain
in 10 on 12pt 'Monophoto' Times Mathematics Series 569
by Page Bros (Norwich) Ltd, Norwich

CONTENTS

INTRODUCTION

This introduction to statistical mechanics is presented for first- or second-year graduate students. It is assumed that the reader has an at least qualitative knowledge of quantum mechanics and that he is familar with thermodynamics.

Quantum mechanics can be based on essentially one fundamental postulate, which is the Schrödinger equation. Similarly, statistical mechanics can be developed from one basic equation, the master equation, from which one can derive, in principle, all the thermodynamic properties of systems containing a large number of particles.

It is possible to deduce the master equation from the Schrödinger equation. This was first done by W. Pauli, and his proof, which is quite simple, is given in the appendix. The reader who is familar with quantum-mechanical perturbation theory, will be able to follow the proof without difficulty. More general derivations exist (for example, the derivation given by L. Van Hove), but they are too involved mathematically to be given in an introductory text.

In order to avoid a discussion of purely quantum-mechanical questions, the existence of the master equation is simply postulated, which amounts to replacing a postulate of quantum mechanics by a postulate of statistical mechanics.

The advantages gained by starting from the master equation are numerous. One should recall that the usual formulations of statistical mechanics which avoid using the master equation, make use of several postulates. To mention merely a few, there is the postulate of equal *a priori* probabilities, the definition (which in fact amounts to a postulate) of the state of thermodynamic equilibrium, the postulate of G. W. Gibbs which states that the thermodynamic properties of a system can be obtained by calculating averages over an ensemble of thermodynamic systems which are copies of the original system. If instead one starts from the master equation, the Gibbs postulate can be forgotten, and the other postulates come out as properties of the solution of this equation.

Another advantage of formulating statistical mechanics using the master equation is that this equation provides a general framework within which both systems that are in thermodynamic equilibrium, and systems approaching equilibrium, can be uniformly studied.

Equilibrium and non-equilibrium thermodynamics thus become complementary aspects of a similar subject.

For essentially physical reasons, this book is concerned almost entirely with systems in thermodynamic equilibrium.

In the first chapter we shall obtain properties of physical systems in thermodynamic equilibrium using a method expounded in part in an article by H. Primakoff and A. Sher [15].

In the second chapter we shall study perfect gases of bosons, fermions, and classical Boltzmann particles. It seemed useful to discuss the Darwin–Fowler method if only to show in a concrete case the physical equivalence of the canonical and grand canonical distributions.

The third chapter is devoted to the imperfect classical gas. Here the simple and direct method of N. Van Kampen is followed, rather than the classical method of Ursell, which, in any event can be found in most textbooks listed in the bibliography. The Van Kampen method allows the partition function to be expressed directly in terms of the density of the gas, and thus the equation of state of imperfect gases is obtained immediately.

In the last chapter it was necessary to make a choice of subjects. Solids form such a vast field for study that the choice was perforce arbitrary.

I am grateful to Professors H. Primakoff and B. Jancovici for their numerous comments and suggestions.

CHAPTER 1

The Distributions of Statistical Mechanics

1.1 Introduction

The aim of statistical mechanics is the study of the properties of physical systems containing a very large number of particles (of the order of 10^{23}). Here we shall limit ourselves to the study of systems that are in thermodynamic equilibrium. The reason for this is that, in fact, most physical systems either are in thermodynamic equilibrium, or else they tend to reach very rapidly a state of equilibrium.

We shall study first the simplest physical systems (perfect gases) and then the more complicated ones (imperfect gases, solids, etc.). We shall obtain the equation of state and the various thermodynamic functions for each system, and examine some specific phenomena, for instance those connected with phase transitions.

Clearly, it would be neither possible nor even desirable to study the individual properties of each of the enormous number of particles comprising a physical system. Even within the framework of classical mechanics where it is in principle possible to know precisely the coordinates and momenta of each particle at any given time, one would need very powerful computers to deduce from this data those global properties of the system that are of physical interest.

Consequently, we shall be interested in the statistical properties of systems. We shall see that, starting from essentially one basic postulate, one is able to deduce very general results that agree with experiments.

The reader may well ask: why is it necessary to introduce even a single postulate? Should not the laws of statistical mechanics follow directly from the laws of quantum mechanics which describe all known natural phenomena? The answer is that they do. However, it is not an easy matter to deduce the laws of statistical mechanics from the laws of quantum mechanics, and besides, it has so far not been possible to avoid introducing an additional postulate in the course of the derivations.*

In order to avoid a discussion that belongs more to quantum mechanics than to statistical mechanics, we shall state the basic postulate of statistical mechanics, try to justify it by rather intuitive arguments, and see whether the results that follow from it are or are not in agreement with experiments. If they are, then we will have good reason to believe that the theory is well founded.

1.2 The Master Equation and its Solution

Consider a completely isolated system made up of two sub-systems A and B. We shall call A the 'system', B the 'thermostat' or the 'heat reservoir' and the combined sub-systems $A + B$ the 'supersystem'. We shall suppose that there exists only a weak interaction between the system A and the thermostat B.

More precisely, if H_A, H_B and H are, respectively, the Hamiltonians corresponding to the system, to the thermostat and to their interaction, the supposition we are making is that the coupling between the system and the thermostat is such that the expectation values of the Hamiltonians obey the conditions†

$$\bar{H} \ll \bar{H}_A \quad \text{and} \quad \bar{H} \ll \bar{H}_B. \tag{1.1}$$

These conditions are quite reasonable for the following reason. The interaction between A and B will result essentially from a Coulomb interaction between the constituents of A and those of B. As bulk matter is electrically neutral, only those constituents of A that are close to those of B will have a non-negligible interaction. If we imagine that A is surrounded by B, then only those particles that are on the surface common to both A and B will have a non-negligible interaction. The ratio between $\bar{H}$ and $\bar{H}_A + \bar{H}_B$ will be proportional to the surface-to-volume ratio and since, for a given density, the number of particles is proportional to the volume, we will have

$$\frac{\bar{H}}{\bar{H}_A + \bar{H}_B} \sim \frac{N_A^{\frac{2}{3}}}{N_A + N_B} \sim \frac{1}{N_A^{\frac{1}{3}}} \ll 1,$$

where N_A and N_B are the number of particles in A and in B.

* The postulate is the so-called random-phase approximation for the wave function of the initial state.

† A bar over a quantity will indicate that we are taking its expectation value. The way in which this done will be shown later on.

Physically the effect of the interaction will be the following. Suppose that A and B can each be in any one of a number of accessible states, before the interaction between A and B is turned on. Then, when the interaction is switched on, both A and B will be able to make transitions to any of their respective accessible states, but the states themselves will not be perturbed.

We shall eventually suppose that the system and the thermostat, because of their mutual interaction, can exchange energy and particles, although at times it will be sufficient to consider only energy exchange.

Since the supersystem is isolated, the effect of the interaction will be to modify the distribution of the particles and of the energy between the subsystems, but both the total energy and the total number of particles of $A + B$ remain constant.

Let now p_i be the proability that the supersystem $A + B$ is in the state i. The time variation of this probability can be written as follows:

$$dp_i(t)/dt = \sum_j [H_{ij} p_j(t) - H_{ji} p_i(t)], \qquad (t > 0). \qquad (1.2)$$

The first term in parenthesis describes transitions from the state j to the state i and the second term describes the inverse transitions. The coefficients H_{ij} and H_{ji} are the probabilities of the occurrence of the transitions from the state i to the state j and from the state j to the state i, respectively. A summation over all accessible states j is effected.*

The fact that the terms on the RHS of (1.2) contain the factors p_j and p_i should require no explanation.

Now it can be shown in quantum mechanics that the coefficients H are time-independent, symmetrical, and non-negative,†

$$H_{ij} = H_{ji} \geqslant 0 \qquad (1.3)$$

In fact, since $A + B$ is isolated, the states i and j will have the same energy, to within a non-measurable uncertainty.

Using (1.3), (1.2) can be written

$$dp_i(t)/dt = \sum_j H_{ij}[p_j(t) - p_i(t)], \qquad i = (1, 2, \ldots, n), t > 0. \ (1.4)$$

* See the appendix for a derivation of this equation.

† H_{ij} is in fact proportional to the square of the modulus of the matrix element of H between the eigenstates i and j of $H_A + H_B$ (see the appendix).

The above equation is called the *master equation.* It serves as a starting point not only for the study of systems that are in thermodynamic equilibrium, but also for the study of irreversible processes. For the reasons that were mentioned in the introduction, we shall not enter into a detailed discussion of the master equation but shall accept it as the fundamental postulate of statistical mechanics.

We shall solve this equation and deduce the properties of its solution in the limit as $t \to \infty$. This limit will correspond to the final state of thermodynamic equilibrium of the system. Put

$$h_i = \sum_{j \neq i} H_{ij}. \tag{1.5}$$

Equation (1.4) becomes

$$dp_i(t)/dt = \sum_j H_{ij} p_j(t) - h_i p_i(t), \qquad (i = 1, 2, \ldots, n). \tag{1.6}$$

The condition that the probabilities be normalizable at all times is automatically satisfied since one has

$$\sum_i dp_i(t)/dt = \sum_{i,j} H_{ij}[p_j(t) - p_i(t)] = 0.$$

Thus

$$\sum_i p_i(t) = \text{constant}.$$

This constant can be taken equal to unity.

To solve the system of equations (1.6), let us introduce the Laplace transform of $p_i(t)$:

$$g_i(\gamma) = \int_0^\infty e^{-\gamma t} p_i(t)\, dt.$$

The Laplace transform of dp_i/dt is, by a partial integration,

$$\int_0^\infty e^{-\gamma t} \frac{dp_i(t)}{dt}\, dt = -p_i(0) + \gamma g_i(\gamma).$$

It follows that the Laplace transform of the system of equations (1.6) is

$$(\gamma + h_i) g_i(\gamma) - \sum_j H_{ij} g_j(\gamma) = p_i(0). \tag{1.7}$$

If one introduces the matrix M where

$$M = \begin{vmatrix} -h_i & H_{12} & H_{13} \dots H_{1n} \\ H_{21} & -h_2 & H_{23} \dots H_{2n} \\ \cdot & & \\ \cdot & & \\ \cdot & & \\ H_{n1} & H_{n2} & H_{n3} \dots -h_n \end{vmatrix}$$

then the solution of (1.7) can be written

$$g_i(\gamma) = \frac{\mathscr{P}_i(n-1, \gamma)}{\det[\gamma I - M]}, \tag{1.8}$$

where I is the unit matrix and $\mathscr{P}_i(n-1, \gamma)$ is a polynomial of degree $n-1$ which we need not write down explicitly.

The matrix M has a number of important properties which we shall list below:

(i) Since M is real and symmetric (*cf.* 1.4), its eigenvalues are all real.

(ii) Consider the matrix $(\gamma I - M)$:

$$(\gamma I - M) = \begin{vmatrix} \gamma + h_1 & -H_{12} & -H_{13} \dots . -H_{1n} \\ -H_{21} & \gamma + h_2 & -H_{23} \dots -H_{2n} \\ \vdots & & \\ -H_{n1} & -H_{n2} & -H_{n3} \dots . \gamma + h_n \end{vmatrix}.$$

If one adds to each element of the first column the elements of all the other columns and of the corresponding row, one obtains, since the value of a determinant is unchanged by these operations:

$$\det(\gamma I - M) = \begin{vmatrix} \gamma & -H_{12} & -H_{13} \dots -H_{1n} \\ \gamma & \gamma + h_2 & -H_{23} \dots -H_{2n} \\ \vdots & & \\ \gamma & -H_{n2} & -H_{n3} \dots \gamma + h_n \end{vmatrix}.$$

which shows that $\gamma = 0$ is an eigenvalue of M.

(iii) Let x be an arbitrary column vector

$$x = \begin{pmatrix} x_1 \\ x_2 \\ \vdots \\ x_n \end{pmatrix}$$

with real components $x_1, x_2, \ldots, x_n$, and let x^T be the transposed (row) vector of x. It is easy to verify that, because of (1.3) and (1.5), the quadratic form associated with the matrix M can be written

$$x^T M x = -\sum_{i,j} H_{ij}(x_i - x_j)^2. \tag{1.9}$$

Since H_{ij} is non-negative, the quadratic form (1.9) is negative semi-definite*:

$$x^T M x \leqslant 0. \tag{1.10}$$

Furthermore, since M is real and symmetric, $x^T M x$ can be diagonalized by an orthogonal transformation O. The transformation of coordinates

$$y = O^T x$$

brings (1.9) to its diagonal form

$$x^T M x = \sum_i \gamma_i y_i^2. \tag{1.11}$$

The quantities γ_i $(i = 0, 1, \ldots, n-1)$ are the eigenvalues of M and y_i $(i = 1, 2, \ldots, n)$ are the components of y.

The quadratic form (1.11) can satisfy (1.10) if, and only if, the eigenvalues γ_i $(i = 0, 1, \ldots, n-1)$ are non-positive. This result is important because it will ensure that the probabilities are normalizable at all times.

To simplify the discussion, we shall suppose that the eigenvalues γ_i $(i = 0, 1, \ldots, n-1)$ are non-degenerate. The generalization of the discussion to the case where there are degeneracies presents no real difficulties.

Because of the properties (i), (ii), and (iii) of M, the eigenvalues of M can be arranged in a sequence

* The quadratic form $x^T M x$ is said to be negative semidefinite if the condition (1.10) is satisfied for any real vector x.

$$0 = \gamma_0 < |\gamma_1| < |\gamma_2| < \ldots < |\gamma_{n-1}|.$$

The solution (1.8) can be put in the form

$$g_i(\gamma) = \frac{\mathscr{P}(n-1, \gamma)}{\gamma(\gamma+|\gamma_1|)(\gamma+|\gamma_2|)\ldots(\gamma+|\gamma_{n-1}|)},$$

which can be written

$$g_i(\gamma) = \sum_{\lambda=0}^{n-1} \frac{A_{i\lambda}}{(\gamma+|\gamma_\lambda|)}; \qquad \gamma_0 = 0, \tag{1.12}$$

where

$$A_{i\lambda} = \frac{\mathscr{P}(n-1, -|\gamma_\lambda|)}{-|\gamma_\lambda|(|\gamma_1|-|\gamma_\lambda|)\ldots(|\gamma_{\lambda-1}|-|\gamma_\lambda|)(|\gamma_{\lambda+1}|-|\gamma_\lambda|)\ldots(|\gamma_{n-1}|-|\gamma_\lambda|)}.$$

By using the inversion formula for Laplace transforms (*)

$$p_i(t) = \frac{1}{2\pi i}\int_{c-i\infty}^{c+i\infty} d\gamma\, e^{\gamma t}\, g_i(\gamma) \qquad (c > c_0),$$

and (1.12), one finally obtains the solution of the system of equations (1.6):

$$p_i(t) = A_{i0} + \sum_{\lambda=1}^{n-1} A_{i\lambda}\, e^{-|\gamma_\lambda| t}. \tag{1.13}$$

From (1.13) it is easily seen how thermodynamic equilibrium is reached. The constants $|\gamma_1|^{-1}$, $|\gamma_2|^{-1}$, ..., $|\gamma_{n-1}|^{-1}$ can be interpreted as 'relaxation times'. For times t that are large compared to the largest relaxation time

$$t \gg |\gamma_1|^{-1},$$

the system reaches equilibrium, since for such times $p_i(t)$ tends towards a constant value A_{io},

$$p_i(t) \to A_{i0} \qquad \text{for } t \gg |\gamma_1|^{-1}.$$

This a consequence of the fact that the eigenvalues of M are non-

* The numbers c and c_0 are such that when $c > c_0$ the integral

$$\int_0^\infty |p_i(x)|^2\, e^{-2cx}\, dx$$

is finite.

positive, as they should be in order that the solution $p_i(t)$ of the master equation be normalizable at all t, including $t \to \infty$.

1.3 Completely Isolated System; Microcanonical Distribution

Consider the supersystem $A + B$, completely isolated from the rest of the universe. We shall specify this supersystem by its energy $\mathscr{E}$ defined up to an uncertainty $\Delta\mathscr{E}$, and by the number of particles $\mathscr{N}$ that it contains.

The solution (1.13) of the master equation gives, for times which are large compared to the largest relaxation time $|\gamma_1|^{-1}$ (we shall write $t \to \infty$),

$$p_i(\infty) = A_{0i}. \tag{1.14}$$

In other words, the probability that a state is occupied tends to a constant for sufficiently large times.

Evaluating (1.4) when $t \to \infty$ and using (1.14), one obtains

$$0 = \sum_j H_{ij}(A_{0j} - A_{0i}).$$

Since H_{ij} is non-negative (*cf*. 1.3) one has

$$A_{0i} = A_{0j} = 1/\Omega(\mathscr{E}, \mathscr{N}), \tag{1.15}$$

where $\Omega(\mathscr{E}, \mathscr{N})$ is the total number of accessible states of $A+B$ which have an energy between $\mathscr{E}$ and $\mathscr{E} + \Delta\mathscr{E}$ and a number of particles $\mathscr{N}$.

The relations (1.14) and (1.15) yield

$$p_i(\infty) = 1/\Omega(\mathscr{E}, \mathscr{N}) \tag{1.16}$$

for all states i. Hence, asymptotically, the probabilities $p_i(\infty)$ no longer depend upon their initial values.

We have obtained the following results. After a sufficiently long time ($t \gg |\gamma_1|^{-1}$), the occupation numbers (probabilities) of all the states of an isolated system become equal. These probabilities are no longer time-dependent. The probability distribution has attained its ultimate value, and the state of thermodynamic equilibrium is reached.

The distribution (1.16) is called a *microcanonical* distribution.

1.4 System in Thermal Equilibrium with a Thermostat: Canonical and Grand Canonical Distributions

In the preceding section, we considered the probability distribution of states of a completely isolated system. We now wish to consider the probability distribution of states of the two sub-systems A and B and, more specifically, the probability distribution of states when the two sub-systems are in thermodynamic equilibrium.

Let $p_{i_A, j_B}(t)$ denote the joint probability that at time t, the system A is in a state i and the thermostat B in a state j, these states being characterized by some set of quantum numbers which still need to be specified. The probability $p_{i_A}(t)$ that at time t, the system A is in a state i, whatever the state B happens to be in is

$$p_{i_A}(t) = \sum_j p_{i_A, j_B}(t). \tag{1.17}$$

Similarly, the probability $p_{i_B}(t)$ that at time t, the thermostat B is in a state j, independently of the state in which A is, is given by

$$p_{j_B}(t) = \sum_i p_{i_A, j_B}(t).$$

We shall assume that the system A and the thermostat B can exchange energy and particles. Let then n_A and n_B denote the number of particles which at a given time are in the systems A and B, respectively, and let $\mathcal{N}$ be the total number of particles of the isolated supersystem. The additivity of these numbers gives

$$\mathcal{N} = n_A + n_B.$$

Analogously, let $E_{i_A}^{n_A}$ be the energy of A, when this system is in a state i which contains n_A particles, and let $E_{i_B}^{n_B}$ be the corresponding quantity for the thermostate B. If $\mathscr{E}$ is the energy of the isolated supersystem $A+B$, one has

$$\mathscr{E} = E_{i_A}^{n_A} + E_{j_B}^{n_B}.$$

The probability that for $t \to \infty$, the state i of system A has an energy $E_{i_A}^{n_A}$ and n_A particles is (*cf.* 1.17)

$$\begin{aligned} p_{i_A} &= p_{i_A}(E_{i_A}^{n_A}, n_A) \\ \mathscr{E} &= \sum_j p_{i_A, j_B}(\infty) \\ &= \frac{\Omega_B}{\Omega(\mathscr{E}, \mathcal{N})} (\mathscr{E} - E_{i_A}^{n_A}, \mathcal{N} - n_A), \end{aligned} \tag{1.18}$$

where $\Omega_B(\mathscr{E} - E_{i_A}^{n_A}, \mathscr{N} - n_A)$ is the total number of accessible states of B that have an energy $\mathscr{E} - E_{i_A}^{n_A}$ and a number of particles $\mathscr{N} - n_A$. In order to obtain (1.18) we used the fact that $A+B$ is completely isolated and thus that $p_{i_A, j_B}(\infty)$ has a value given by the micro-canonical distribution (*cf.* 1.16).

Similarly, the probability that for $t \to \infty$ the state j of the thermostat B has an energy $E_{j_B}^{n_A}$ and a number of particles n_B is

$$p_{j_B}(\infty) = p_{j_B}(E_{j_B}^{n_B}, n_{j_B})$$
$$= \sum_i p_{i_A, j_B}(\infty)$$
$$= \frac{\Omega_A}{\Omega(\mathscr{E}, \mathscr{N})}(\mathscr{E} - E_{j_B}^{n_A}, \mathscr{N} - n_B),$$

where $\Omega_A(\mathscr{E} - E_{j_B}^{n_B}, \mathscr{N} - n_B)$ is the number of accessible states of A that have an energy $\mathscr{E} - E_{j_B}^{n_B}$ and a number of particles $\mathscr{N} - n_B$.

If the energy $\mathscr{E}$ and the number of particles $\mathscr{N}$ of the super-system are large compared to the energy $E_{i_A}^{n_A}$ and the number of particles n_A of the system A,

$$E_{i_A}^{n_A} \ll \mathscr{E},$$
$$n_A \ll \mathscr{N}, \tag{1.19}$$

one can expand the numerator of (1.18) in a double Taylor series. Instead of expanding Ω_B directly, we shall expand the more slowly-varying function $\ln \Omega_B$:

$$p_{i_A} = \frac{\exp[\ln \Omega_B(\mathscr{E} - E_{i_A}^{n_A}, \mathscr{N} - n_A)]}{\Omega(\mathscr{E}, \mathscr{N})}$$
$$= \frac{1}{\Omega(\mathscr{E}, \mathscr{N})} \exp[\ln \Omega_B(\mathscr{E}, \mathscr{N}) + \alpha n_A + \beta E_{i_A}^{n_A} + \ldots] \tag{1.20}$$
$$\triangleq \frac{\Omega_B(\mathscr{E}, \mathscr{N})}{\Omega(\mathscr{E}, \mathscr{N})} \exp(\alpha n_A + \beta E_{i_A}^{n_A} + \ldots),$$

where

$$\alpha \equiv -\frac{\partial \ln \Omega_B(\mathscr{E}, \mathscr{N})}{\partial \mathscr{N}}, \quad \beta \equiv -\frac{\partial \ln \Omega_B(\mathscr{E}, \mathscr{N})}{\partial \mathscr{E}}. \tag{1.21}$$

The normalization of the probabilities gives

$$\sum_i p_{i_A} = 1 = \frac{\Omega_B(\mathscr{E}, \mathscr{N})}{\Omega(\mathscr{E}, \mathscr{N})} \sum_{i, n_A} \exp(\alpha n_A + \beta E_{i_A}^{n_A})$$

Putting

$$z = \exp \alpha \tag{1.22}$$

and

$$Q_A = \sum_{n_A} \sum_i z^{n_A} \exp(\beta E_{i_A}^{n_A}) \tag{1.23}$$

one finds

$$p_{i_A} = z^{n_A} \exp(\beta E_{i_A}^{n_A})/Q_A \tag{1.24}$$

The distribution (1.24) is called a *grand canonical* distribution. The function (1.23) is called the *grand partition function* and the quantity z defined by (1.22) is called the *activity*.

Notice that in (1.20) it is the function Ω_B, the number of accessible states of the thermostat, that is expanded in a Taylor series. This implies that the energy spectrum of the thermostat is continuous, which in turn means that its volume is sufficiently large. This condition, together with the conditions (1.19), imply that the thermostat is a system of macroscopic dimensions and contains a large number of particles. There are no restrictions on the system A, aside from the conditions (1.19). Thus, A can be a macroscopic system or, on the contrary, it could contain no more than one molecule.

Suppose now that the system and the thermostat can exchange energy but can no longer exchange particles. This means that the number of particles in the system A and in the thermostat B remain constant. The number of states Ω_B of the thermostat will be a function of the energy only and the term proportional to α in the expansion (1.21) will no longer appear. In this case one simply obtains

$$p_{i_A} = \exp(\beta'_A E_{i_A})/q_A \tag{1.25}$$

where

$$q_A = \sum_i \exp(\beta' E_{i_A}) \tag{1.26}$$

and where β' is a constant which remains to be determined.

The distribution (1.25) is called a *canonical* distribution while (1.26) is called the *partition function*.

The partition function and the grand partition function are among the most important functions of statistical mechanics. From them one can, in principle, calculate all the thermodynamic properties of a system. The average value of a quantity F_i is defined as

$$\bar{F} = \sum_i p_i F_i \tag{1.27}$$

where p_i is given by (1.16), (1.24) or (1.25), according as to whether one uses the microcanonical, the grand canonical or the canonical distribution. The microcanonical distribution is seldom used in applications because it is best suited to describe completely isolated systems which rarely occur in nature. It should be clearly understood, however, that the various distributions are different analytical tools which must lead to the same physical results. In practice, one uses the distribution that is best adapted to the problem considered. In chapter 2 we will treat the same problem using the canonical and grand canonical distributions and display the equivalence of the results.

In section 1.6 we shall determine the parameters α, β, β' which appear in (1.22), (1.24), and (1.25).

1.5 Entropy, Temperature, and the First Two Laws of Thermodynamics

As before we let $p_{i_A, j_B}(t)$ denote the joint probability that within the supersystem $A+B$, A is in the state i and B in the state j. The *entropy* of the supersystem will be, by definition:

$$S_{A,B}(t) \equiv -k \sum_i \sum_j p_{i_A, j_B}(t) \ln p_{i_A, j_B}(t), \tag{1.28}$$

where k is Boltzmann's constant ($k = 1{\cdot}3840 \times 10^{-16}$ erg/K) and the summation extends over all possible states of A and of B. The entropies of the two subsystems A and B are given by analogous expressions (*cf*. 1.17)

$$\begin{aligned} S_A(t) &\equiv -k \sum_i p_{i_A}(t) \ln p_{i_A}(t) \\ &\equiv -k \sum_i \sum_j p_{i_A, j_B}(t) \ln \Big[\sum_l p_{i_A, l_B}(t)\Big] \end{aligned} \tag{1.29}$$

$$\begin{aligned} S_B(t) &\equiv -k \sum_j p_{j_B}(t) \ln p_{j_B}(t) \\ &\equiv -k \sum_j \sum_i p_{i_A, j_B}(t) \ln \Big[\sum_m P_{m_A, j_B}(t)\Big] \end{aligned} \tag{1.30}$$

Using the master equation, we shall now demonstrate an important property of $S_{A,B}(t)$ (1.28).

Taking the time derivative of (1.28) one has

$$\frac{d(S_{A,B}(t))}{dt} = -k\left[\sum_{i,j} \frac{dp_{i_A,j_B}}{dt} \ln p_{i_A,j_B} + \sum_{i,j} \frac{dp_{i_A,j_B}}{dt}\right]. \tag{1.1}$$

The second term vanishes since

$$\sum_{i,j} \frac{dp_{i_A,j_B}}{dt} = \frac{d}{dt}\sum_{i,j} p_{i_A,j_B} = \frac{d}{dt}1 = 0. \tag{1.32}$$

The time derivative of p_{i_A,j_B} is given by the master equation. We recall that this equation holds for an isolated system.

In order that the notation be more precise, we shall rewrite (1.4) as follows:

$$\frac{dp_{i_A,j_B}(t)}{dt} = \sum_{i',j'} H_{i_A,j_B;i'_A,j'_B}[p_{i'_A,j'_B} - p_{i_A,j_B}],$$

where $H_{i_A,j_B;i'_A,j'_B}$ describes transitions between the states i of A, j of B and the states i' of A, j' of B. Substituting this expression in (1.31), one obtains, using (1.32):

$$\frac{dS_{A,B}(t)}{dt} = -k\sum_{i,j,i',j'} H_{i_A,j_B;i'_A,j'_B}[p_{i'_A,j'_B} - p_{i_A,j_B}] \ln p_{i_A,j_B} \tag{1.33}$$

or, what is the same, making use of (1.3)

$$\frac{dS_{A,B}(t)}{dt} = -k\sum_{i,j,i',j'} H_{i_A,j_B;i'_A,j'_B}[p_{i_A,j_B} - p_{i'_A,j'_B}] \ln p_{i'_A,j'_B}. \tag{1.34}$$

Taking half the sum of (1.33) and (1.34) one finds:

$$\frac{dS_{A,B}(t)}{dt} = \frac{1}{2}k\sum_{i,j,i',j'} H_{i_A,j_B;i'_A,j'_B}[p_{i_A,j_B} - p_{i'_A,j'_B}] \ln \frac{p_{i_A,j_B}}{p_{i'_A,j'_B}}.$$

But it is obvious that

$$(p_{i_A,j_B} - p_{i'_A,j'_B}) \ln \frac{p_{i_A,j_B}}{p_{i'_A,j'_B}} \geqslant 0$$

in all cases (i.e. when $p_{i_A,j_B} \geqslant p_{i'_A,j'_B}$ or when $p_{i_A,j_B} < p_{i'_A,j'_B}$).

Therefore

$$dS_{A,B}(t)/dt \geqslant 0 \tag{1.35}$$

and the entropy of the isolated supersystem $A+B$ either increases, or is stationary. At equilibrium one has

$$dS_{A,B}/dt = 0, \tag{1.36}$$

and thus $S_{A,B}$ is a constant.

It is important to realize that the entropies of the two non-isolated sub-systems A and B do not have the property (1.35) of increasing in time. A calculation analogous to the one that led to this result, shows that S_A and S_B are not necessarily monotonically increasing functions of the time.

Another result which follows immediately from the definitions (1.28), (1.29), and (1.30) is that, in general, entropy is not an additive quantity, i.e.

$$S_{A,B} \neq S_A + S_B. \tag{1.37}$$

The situation changes radically however, if, after a sufficiently long time, the system and the thermostat become statistically independent in the sense that

$$p_{i_A} p_{j_B} = p_{i_A, j_B}. \tag{1.38}$$

If this happens, then one obtains

$$\begin{aligned} S_{A,B}(\mathscr{E}) &= -k \sum_{i,j} p_{i_A}(\infty)\, p_{j_B}(\infty) \ln [p_{i_A}(\infty\, p_{j_B}(\infty)] \\ &= -k \sum_i p_{i_A}(\infty) \ln p_{i_A}(\infty) - k \sum_j p_{j_B}(\infty) \ln p_{j_B}(\infty) \\ &= S_A(E_A) + S_B(E_B). \end{aligned} \tag{1.39}$$

Entropy thus becomes additive. The assumption (1.38) will henceforth be made for all systems that are in thermodynamic equilibrium.

At equilibrium it follows from this property of the additivity of the entropy (since $E_B = \mathscr{E} - E_A$)

$$\frac{\partial S_{A,B}}{\partial E_A} = 0 = \frac{\partial S_A(E_A)}{\partial E_A} - \frac{\partial S_B(E_B)}{\partial E_B}$$

whence

$$\frac{\partial S_A(E_A)}{\partial E_A} = \frac{\partial S_B(E_B)}{\partial E_B}. \tag{1.40}$$

Hence when two systems are in thermodynamic equilibrium, the derivatives of the entropy of each system with respect to the energy are equal. Naturally the equilibrium conditions between A and B will be perturbed if the volume of either A or B is changed by an external constraint. Taking this fact into account, one can define the absolute temperature T of a system by the relation

$$\left(\frac{\partial S}{\partial E}\right)_V = \frac{1}{T}. \tag{1.41}$$

The relation (1.40) then shows that two systems that are in thermodynamic equilibrium have the same temperature.

Before closing this section we recall that the first two laws of thermodynamics are embodied in the relation

$$dE = T\,dS - P\,dV \tag{1.42}$$

which is equivalent to (1.41) together with the relation

$$P = T\left(\frac{\partial S}{\partial V}\right)_E. \tag{1.43}$$

1.6 Determination of Parameters in the Partition and Grand Partition Functions

In the definitions (1.26) and (1.23) of the partition function and of the grand partition function, there appeared parameters α, β and β' which we shall now relate to thermodynamic quantities.

From here on, we shall focus our attention on the system A exclusively. We shall therefore omit all references to A, which will be implicit.

From (1.29), (1.26), and (1.27), one has

$$\begin{aligned} S &= -k\sum_i \frac{\exp(\beta' E_i)}{q}[\beta' E_i - \ln q] \\ &= -k\beta' E + k\ln q, \end{aligned} \tag{1.44}$$

where E is the mean energy of the system.* Taking the differential of S, one finds

* Often we will omit bars over quantities that represent mean values when the significance is obvious.

$$dS = -kE\,d\beta' - k\beta'\,dE + \frac{k}{q}[\sum_i E_i \exp(\beta' E_i]\,d\beta'$$

$$= -kE\,d\beta' - k\beta'\,dE + kE\,d\beta'$$

$$= -k\beta'\,dE. \tag{1.45}$$

A comparison of (1.45) and (1.41) leads to

$$\left(\frac{\partial S}{\partial E}\right)_{V,N} = \frac{1}{T} = -k\beta',$$

whence

$$\beta' = -1/kT. \tag{1.46}$$

Similarly, from (1.44) and (1.46) one has

$$-\ln q = \frac{E-TS}{kT} \equiv \frac{A}{kT}, \tag{1.47}$$

where A is the Helmholtz free energy. Thus

$$q = e^{-A/kT}. \tag{1.48}$$

This very important relation connects the partition function to thermodynamics.

Summarizing, the canonical distribution is given by

$$p_i = \exp(-E_i/kT)/q$$

where

$$q = \exp(-A/kT) = \sum_i \exp(-E_i/kT).$$

The parameters of the grand canonical distribution are obtained in an analogous fashion. Putting (1.24) into (1.29), one has

$$S = -k\sum_{i,n} \frac{\exp(\beta E_i^n)}{Q}[\beta E_i^n + n \ln z - \ln Q]$$

$$= -k\beta E - k\alpha N + k \ln Q, \tag{1.49}$$

where E and N are the average energy and the average number of particles of the system. Taking the differential of (1.49),

$$dS = -k\beta\,dE - kE\,d\beta - k\alpha\,dN - kN\,d\alpha + kd\,(\ln Q). \tag{1.50}$$

But, using (1.23), one finds

$$d(\ln Q) = \frac{1}{Q}\left(\frac{\partial Q}{\partial \beta}\right)_{\alpha, V} d\beta + \frac{1}{Q}\left(\frac{\partial Q}{\partial \alpha}\right)_{\beta, V} d\alpha$$

$$= E\,d\beta + N\,d\alpha. \tag{1.51}$$

Combining (1.50) and (1.51) leads to

$$dS = -k\beta\,dE - k\alpha\,dN. \tag{1.52}$$

On the other hand, one has the following thermodynamic relations:

$$dS = \frac{dQ}{T} = \frac{dE + P\,dV - \mu\,dN}{T}, \tag{1.53}$$

where μ is the chemical potential. From this relation one obtains:

$$\left(\frac{\partial S}{\partial E}\right)_{V,N} = \frac{1}{T}, \qquad \left(\frac{\partial S}{\partial N}\right)_{V,E} = -\frac{\mu}{T}.$$

Thus, comparing the above with (1.52),

$$\alpha \equiv \frac{\mu}{kT}, \qquad \beta \equiv -\frac{1}{kT}. \tag{1.54}$$

Finally, using (1.49) and (1.54),

$$\ln Q = \frac{ST - E + \mu N}{kT} = \frac{PV}{kT}, \tag{1.55}$$

where the last equality follows from thermodynamics.

Let us summarize the important formulas of the grand canonical distribution:

$$Q = \exp(PV/kT) = \sum_{i,n} z^n \exp(-E_i^n/kT). \tag{1.56}$$

$$z = \exp(\mu/kT). \tag{1.57}$$

$$p_i^n = z^n \exp(-E_i^n/kT)/Q. \tag{1.58}$$

1.7 Fluctuations in the Canonical Distribution

The statistical treatment given so far can be considered valid only if the fluctuations of the various thermodynamic quantities from their mean values, are negligible. To illustrate a typical calculation, we shall calculate the energy fluctuations in the canonical distribution.

Referring to (1.26), one has

$$\left(\frac{\partial \ln q}{\partial \beta}\right)_{V,N} = \frac{1}{q}\sum_i E_i e^{\beta' E_i} = E. \tag{1.59}$$

Similarly

$$\overline{E^2} \equiv \frac{1}{q}\sum_i E_i^2 e^{\beta' E_i} = \frac{1}{q}\frac{\partial^2 q}{\partial \beta'^2}. \tag{1.60}$$

Thus, using (1.59) and (1.60)

$$\overline{(\Delta E)^2} = \overline{E^2} - E^2 = \frac{1}{q}\frac{\partial^2 q}{\partial \beta'^2} - \left(\frac{1}{q}\frac{\partial q}{\partial \beta'}\right)^2$$

$$= \frac{\partial E}{\partial \beta'} = kT^2\frac{\partial E}{\partial T} = kT^2 c_V \tag{1.61}$$

where c_V is the specific heat of the system at constant volume. Thus the energy fluctuations in the canonical distribution are proportional to the specific heat.

Example For a perfect classical gas, one has

$$E = \tfrac{3}{2}NkT.$$

(This formula will be proved later.) Thus the root mean square fluctuation in energy is given by

$$\sqrt{\frac{\overline{(\Delta E)^2}}{E^2}} = \sqrt{\frac{2}{3N}}$$

and the fluctuations become negligible when $N \gg 1$.

1.8 Fluctuations in the Grand Canonical Distribution

The energy fluctuations in the grand canonical distribution are calculated as in the preceding section. We shall here calculate the fluctuations in the particle number. One has

$$N = \frac{1}{Q}\sum_{i,n} nz^n \exp(E_i^n/kT) = z\frac{\partial}{\partial z}\ln Q \tag{1.62}$$

from which it follows that

$$z\frac{\partial}{\partial z}N = -\frac{1}{Q^2}\left[\sum_{i,n} nz^n \exp(-E_i^n/kT)\right]^2 + \frac{1}{Q}\left[\sum_{i,n} n^2 z^n \exp(-E_n^i/kT)\right]$$

$$= -N^2 + \overline{N^2} \equiv (\Delta N)^2$$

On the other hand

$$\left(\frac{\partial N}{\partial P}\right)_T = \left(\frac{\partial N}{\partial z}\right)_T \left(\frac{\partial z}{\partial P}\right)$$

$$\equiv \left(z\frac{\partial N}{\partial z}\right)_T \left(\frac{1}{z}\cdot\frac{\partial z}{\partial P}\right)_T$$

$$= \left(z\frac{\partial N}{\partial z}\right)_T \left(\frac{\partial \ln z}{\partial P}\right)_T$$

With (1.57) and the thermodynamic relation

$$N\mu = E - TS + PV$$

we get

$$\left(\frac{\partial N}{\partial P}\right)_T = (\Delta N)^2 \frac{V}{NkT}. \tag{1.63}$$

Introducing the density $\rho = N/V$, (1.63) can be written

$$(\Delta N)^2 = NkT\left(\frac{\partial \rho}{\partial P}\right)_T. \tag{1.64}$$

Thus the fluctuations in particle number are proportional to the isothermal compressibility of the gas.

Example From the perfect gas law $PV = NkT$ one obtains

$$(\Delta N)^2 = N$$

Thus the root mean square fluctuations

$$\sqrt{\frac{(\Delta N)^2}{N^2}} = \frac{1}{\sqrt{N}}$$

are again negligible when $N \gg 1$.

1.9 Classical and Quantal Systems: the Thermodynamic Limit

We have written the partition function and the grand partition function as sums over discrete states i. Thus we have tacitly assumed that we were dealing with a quantized system. For a classical system, where the states form a continuous rather than a discrete spectrum, these sums must be replaced by integrals. The question is, how does one go about it?

For a classical system it is convenient to introduce the notion of a phase space. Each molecule of a gas can be dynamically specified by a generalized coordinate $\vec{q}$ and by its conjugate momentum $\vec{p}$, i.e. by the six components q_k and p_k (k = 1, 2, 3). The N molecules of a gas can therefore be specified by the 6N components

$$q_1, q_2, \ldots, q_{3N}; p_1, p_2, \ldots, p_{3N}. \tag{1.65}$$

If one now considers a Cartesian coordinate system in $6N$ dimensions bearing along its axes the $6N$ components (1.65), the dynamical state of the system of N molecules will be specified at each instant by a point in this $6N$ dimensional space, and the evolution of the system is time will be represented by a trajectory in the space. The space is called a *phase space.* Rather than speak of the 'number of states' one now speaks of the *volume* that the system occupies in phase space

$$dv = dp_1\, dp_2 \ldots dp_{3N}\, dq_1\, dq_2 \ldots dq_{3N} \tag{1.66}$$

and the notion of the number of states of a quantal system is replaced, for a classical system, by the notion of *accessible volume in phase space.*

It is important to note that the notion of phase space is not used in this context in quantum mechanics, for, on account of the indeterminacy principle, it is impossible to simultaneously specify the coordinate and the conjugate momentum of a particle.

In order to be able to make a transition from a sum over discrete states to an integral over continuous states, one needs to know what volume in phase space corresponds to a discrete quantum state. We shall establish the important results that *a discrete quantum state occupies a volume equal to* h^{3N}, where h is Planck's constant (h = 6·62 × 10Js).

Let us prove this result in the particular case of a particle moving freely within a cube of edge L. The Schrödinger equation

$$-\frac{\hbar^2}{2m}\left(\frac{\partial^2\psi}{\partial x_1^2}+\frac{\partial^2\psi}{\partial x_2^2}+\frac{\partial^2\psi}{\partial x_3^3}\right) = E\psi$$

has a solution

$$\psi = \frac{1}{L^{\frac{3}{2}}}\exp[i(p_1 x_1+p_2 x_2+p_3 x_3)/\hbar] \tag{1.67}$$

The wave function must satisfy appropriate boundary conditions, and we shall choose the periodic boundary conditions

$$\psi(x_j+L) = \psi(x_j) \qquad (j = 1, 2, 3)$$

These conditions, applied to (1.67), yield the relations

$$p_1 = \frac{h}{L}n_1; \quad p_2 = \frac{h}{L}n_2; \quad p_3 = \frac{h}{L}n_3 \tag{1.68}$$

where n_1, n_2, n_3 are integers. Thus the momenta assume discrete values and an element of volume in phase space will contain as many discrete states as there are points within this volume which satisfy the conditions (1.68). An element of volume d^3p will therefore contain

$$\frac{L^3}{h^3}d^3p = V\frac{d^3p}{h^3}$$

discrete states of particles, where V is the volume of the cube.

Generalizing this result to the case where N particles more freely within an N-dimensional hypercube, one arrives at the conclusion that an element of volume (1.66) contains $V\,d^3p_1, d^3p_2, \ldots, d^3p_3/h^{3N}$ discrete states of N particles where V is the volume of the hypercube

$$V = \int \ldots \int d^3q_1 d^3q_N.$$

This proves the statement made earlier in a particular case. The result, however, is quite general. The reader can find an elegant proof in Dirac's book.

Example The Hamiltonian for an harmonic oscillator of mass m and frequency ω is

$$H = \frac{p^2}{2m}+\frac{m\omega^2}{2}q^2 = E.$$

For a classical oscillator, E will be able to take on a continuous set of values, but if the oscillator is quantized, only the discrete set of values

$$E = \hbar\omega(n+\tfrac{1}{2}), \qquad (n = 0, 1, 2, \ldots)$$

are allowed. In phase space, the oscillator will describe an ellipse

$$\frac{p^2}{2mE}+\frac{m\omega^2}{2E}q^2 = 1$$

and if the energy assumes discrete values, the surface in phase space between the $(n+1)$th and the nth discrete states will be given by the area between the two ellipses

$$\begin{aligned} A_{n+1}-A_n &= \pi\left\{\sqrt{(m\hbar\omega(2n+3))}\sqrt{\left[\frac{\hbar}{m\omega}(2n+3)\right]}\right. \\ &\qquad \left.-\sqrt{[m\hbar\omega(2n+1)]}\sqrt{\left[\frac{\hbar}{m\omega}(2n+1)\right]}\right\} \\ &= 2\pi\hbar = h \end{aligned}$$

This area does not depend upon n and its value is in accord with the result obtained above.

It is now easy to see how one can make a transition from a sum over discrete states to an integral over phase space. One must make the following substitution:

$$\sum_i \to \delta\int\ldots\int\frac{d^3q_1\ldots d^3q_N\, d^3p_1\ldots d^3p_N}{h^{3N}} \tag{1.69}$$

where δ is a normalization constant, which may depend on N, and which will be determined later on.

Let us also note that according to (1.68), the components of the momenta tend to form a continuous set of states when $L \to \infty$. In other words, in the limit of an infinite volume, a sum over states can, generally,* be replaced by an integral with weight

$$V\, d^3p/h^3$$

even if the system is quantized. The limit of an infinite volume implies necessarily that the number of particles in the system must also tend to infinity, but in such a way that the density of the gas N/V

* An exceptional case will be considered in the next chapter.

remains finite. The limit

$$V \to \infty, \quad N \to \infty, \quad N/V < \infty \tag{1.70}$$

is called the *thermodynamic limit*.

EXERCISES

1 Starting from the grand canonical distribution, show that the probability P_N that the system contains N particles is given by the Poisson distribution

$$P_N = \frac{1}{\bar{N}!} e^{-\bar{N}} (\bar{N})^N$$

where $\bar{N}$ is the average number of particles.

2 Find the master equation for system A. Show that it has the same general form as the master equation for the isolated super-system $A+B$, but that the coefficients of proportionality may be time-dependent.

3 The Hamiltonian of a system of N uncoupled harmonic oscillators of mass m is given by the expression

$$H = \frac{1}{2m} \sum_{i=1}^{N} (p_i^2 + mkq_i^2)$$

where k is a constant. Find the solution of Hamiltonian's equations and show by an explicit calculation that the volume in phase space that the oscillators occupy

$$V(t) = \int_{\Omega(t)} \prod_{i=1}^{N} dp_i \, dq_i$$

(where $\Omega(t)$ is an arbitrary surface that encloses the N oscillators at time t) is a Poincaré integral invariant (i.e. $V(t)$ is time-independent). This result is a particular case of a theorem due to Liouville.

4 Consider a system of N particles of spin $\frac{1}{2}$ in a magnetic field. Let E denote the difference between the energies of a single particle when its spin is pointing in the direction of the magnetic field and when it is pointing in a direction opposite to the magnetic field. Show that there is an energy range where the temperature of the system can be negative. Compare the energies of the system in the ranges where the temperature is positive and negative.

CHAPTER 2

The Perfect Gas

2.1 Introduction

A *perfect gas* is, by definition, a gas consisting of atoms or molecules with negligible mutual interactions. Therefore the total energy of each particle of a perfect gas will be due to its kinetic energy alone. The neglect of interactions between particles is, of course, an approximation, but we shall see that it is a very good approximation in the case of rarefied gases. Furthermore, the study of perfect gases is important because it serves as a starting point in the study of imperfect gases.

We shall consider gases consisting of identical particles.* Among identical particles, we shall have to make the distinction between distinguishable particles (described by classical mechanics) and indistinguishable particles (described by quantum mechanics). The indistinguishable particles will further have to be classified according as to whether they are *bosons* or *fermions*.

2.2 Indistinguishable and Distinguishable Particles: Symmetries

Consider a gas consisting of N identical particles. In quantum mechanics, identical particles are regarded as indistinguishable.

By the 'indistinguishability of N particles' one is to understand that no experiment can distinguish one particle from another. Let now $\Psi(1, 2, \ldots, N)$ be the wave function of this system, obeying the Schrödinger equation

$$H\psi(1, 2, \ldots, N) = E\psi(1, 2, \ldots, N).$$

The numbers $1, 2, \ldots, N$ represent the totality of the properties

* The following considerations can easily be extended to the case of a gas consisting of many substances, each substance being composed of identical particles.

which characterize the particles numbered from 1 to N. Of course, this numbering is completely artificial since the particles are indistinguishable. Therefore, in order to remove all traces of this labelling, one introduces a permutation operator P, which permutes some arbitrary sub-group of the N particles. In total there are $N!$ possible permutations. Since the particles are indistinguishable, the Hamiltonian is invariant under any such permutation, and P commutes with H

$$PH = HP.$$

Hence, applying the operator P to the above Schrödinger equation,

$$P\{H\Psi(1,2,\ldots,N)\} = H\{P\Psi(1,2,\ldots,N)\} = E\{P\Psi(1,2,\ldots,N)\},$$

which implies that $P\Psi(1,2,\ldots,N)$ is also an eigenfunction of H corresponding to the energy E. We therefore have $N!$ wave functions, or equivalently $N!$ linear combinations of these wave functions which are possible candidates to describe a system of N identical particles that have an energy E. However it is a fortunate experimental circumstance that there exist in nature only two different categories of particles that are in fact described by two very particular linear combinations of these wave functions. One of these categories contains particles described by a wave function which is *totally symmetric* with respect to the interchange of any pair of particles,

$$\psi_{\text{S.}}(1,2,\ldots,N) = \frac{1}{N!}\sum_P P\psi(1,2,\ldots,N). \tag{2.1}$$

The summation extends over the $N!$ possible permutations of the N particles. The other category of particles contains particles described by a wave function which is *totally antisymmetric* with respect to the interchange of any pair of particles

$$\psi_{\text{A.S.}}(1,2,\ldots,N) = \frac{1}{N!}\sum_P \varepsilon_P\psi(1,2,\ldots,N), \tag{2.2}$$

where $\varepsilon_P = +1$ for an even permutation of $1,2,\ldots,N$,
$\varepsilon_P = -1$ for an odd permutation of $1,2,\ldots,N$.

A system of particles described by a totally symmetric wave function is called a system of *bosons*; one also says that the system obeys *Bose–Einstein statistics*. A system of particles described by a totally antisymmetric wave function is called a system of *fermions*; it is also said that the system obeys *Fermi–Dirac statistics*.

A very important result which emerges from relativistic quantum field theory which, however, we can only state here is the following: all particles with integer spin are bosons, all particles with half-integer spins are fermions. Thus helium nuclei, photons, π and K mesons are bosons whereas electrons, protons, neutrons, μ mesons are fermions.

In the case of free particles, the wave function (2.2) can be written in a more suggestive fashion for, in this case, the wave function $\psi(1, 2, \ldots, N)$ can be resolved into a product of 1-particle wave functions

$$\psi(1,2,\ldots,N) = \psi_i(1)\psi_j(2)\ldots\psi_r(N),$$

where the indices $i, j, \ldots, r$ label the possible energy states of the particles. Using this equation, the totally antisymmetric wave function (2.2) can be written in the form of a determinant

$$\psi_{\text{A.S.}}(1,2,\ldots,N) = \begin{vmatrix} \psi_i(1) & \psi_i(2)\ldots\psi_i(N) \\ \psi_j(1) & \psi_j(2)\ldots\psi_j(N) \\ \vdots & \\ \psi_r(1) & \psi_r(2)\ldots\psi_r(N) \end{vmatrix}$$

It is easy to see that this matrix is totally antisymmetrical; for the interchange of two particles is effected by interchanging two columns of the determinant, and this entails a change of sign. It is also easy to notice that if two particles are in the same state (for example if $\psi_i = \psi_j$), the determinant vanishes. In other words, if a system of particles is described by a totally antisymmetric wave function, no two particles can be in the same state. This result is nothing else but a statement of the Pauli exclusion principle.

The condition that a system of bosons or of fermions has a wave function with specified symmetry properties implies, of course, that there exist certain correlations among the particles, even if the particles are non-interacting.

Classical mechanics describes distinguishable particles. This means that the particles can be numbered and a permutation of the particles will lead to a distinct state.

Taking into account these properties, we shall consider first the Boltzmann or classical gas, and then the boson and fermion gases.

2.3 The Perfect Gas in The Boltzmann Approximation

The total energy of a perfect classical gas consisting of N molecules of mass m is

$$E = \sum_{i=1}^{N} p_i^2/2m.$$

The partition function is given by the expression

$$q = \sum_{p_1, p_2, \ldots, p_N} \exp\left[-\sum_{i=1}^{N} p_i^2/2mkT\right]$$
$$= \left[\sum_{p_1} \exp(-p_1^2/2mkT)\right]\left[\sum_{p_2} \exp(-p_2^2/2mkT)\right]\ldots$$
$$\times \left[\sum_{p_N} \exp(-p_N^2/2mkT)\right] = \left[\sum_{p} \exp(-p^2/2mkT)\right]^N.$$

One can replace the sum above by an integral, using equation (1.69),

$$q = \frac{\delta}{h^{3N}}\left[V \int_0^\infty \exp(-p^2/2mkT)\, 4\pi p^2 \, dp\right]^N.$$

With the result

$$\int_{-\infty}^{\infty} e^{-\alpha x^2} dx = \sqrt{(\pi/\alpha)}$$

one finds

$$q = \delta V^N (2\pi mkT/h^2)^{3N/2} \equiv \delta V^N/\lambda^{3N} \tag{2.3}$$

where

$$\lambda = \lambda(T) \equiv (h^2/2\pi mkT)^{\frac{1}{2}} \tag{2.4}$$

is called the *thermal wavelength* of the particle.

The thermodynamic functions can be derived from (2.3) and the relation

$$-A/kT = \ln q = \ln(\delta V^N/\lambda^{3N}). \tag{2.5}$$

Using equations (1.42) and (1.47) one finds

$$dA = -P\,dV - S\,dT \tag{2.6}$$

and thus

$$P = -(\partial A/\partial V)_T. \tag{2.7}$$

The relations (2.5) and (2.7) immediately lead to the equation of state of the perfect classical gas

$$P = NkT/V. \tag{2.8}$$

The entropy is obtained from (2.6) and (2.5)

$$S = -\left(\frac{\partial A}{\partial T}\right)_v = \frac{3}{2}Nk + k\ln\left(\frac{\delta V^N}{\lambda^{3N}}\right). \tag{2.9}$$

The average energy of the system is obtained from equation (1.47):

$$E = A + TS = \tfrac{3}{2}NKT. \tag{2.10}$$

The normalization constant δ which appears in (2.3), (2.4) and (2.9) must, of course, still be determined. In the next section we shall give the argument which, historically, led to its determination.

2.4 The Gibbs Paradox

Suppose that one has two different gases, one composed of N_1 molecules, the other of N_2 molecules, and suppose that these gases are separated by a membrane. Let V_1 and V_2 be the volumes that these gases occupy. Let $V = V_1 + V_2$ and $N = N_1 + N_2$ be the total volume and the total number of particles. The entropy of each gas is

$$S_i = \tfrac{3}{2}N_i k + k\ln(\delta_i V_i^N/\lambda^{3N_i}).$$

If one removes the membrane, the two gases will mix and the resulting change in entropy will be given by

$$\Delta S = \frac{3}{2}Nk - \frac{3}{2}N_1 k - \frac{3}{2}N_2 k + k\ln\left(\frac{\delta V^N}{\lambda^{3N}}\right)$$
$$- k\ln\left(\frac{\delta_1 V_1^{N_1}}{\lambda^{3N_1}}\right) - k\ln\left(\frac{\delta_2 V_2^{N_2}}{\lambda^{3N_2}}\right)$$
$$= k\ln\left(\frac{\delta V^{N_1}}{\delta_1 V_1^{N_1}}\right) + k\ln\left(\frac{\delta V^{N_2}}{\delta_2 V_2^{N_2}}\right).$$

Originally, the factor δ which, we recall, was introduced as some arbitrary phase space normalization constant, was assumed to be equal to one. This led to a famous paradox discovered by Gibbs; for then one sees immediately that one would have $\Delta S > 0$ even if the gases are identical, i.e. even if, strictly speaking, there is no mixing. It is no wonder that the factor δ had not been introduced at the time of Gibbs, for it is only within the framework of quantum theory that it can be understood.

A quantal gas is composed of indistinguishable particles. Therefore a permutation of the N particles in a given state of the gas cannot alter that state. In the case of a Boltzmann gas, such a permutation would lead to $N!$ different microscopic states of the system corresponding, however, to the *same* macroscopic state. Thus, a quantal gas has $1/N!$ fewer states than the corresponding Boltzmann gas.

Clearly, then, one should take the normalization constant of equation (1.69) to be

$$\delta_i = \frac{1}{N_i!}$$

The partition function for a Boltzmann gas is then

$$q = \frac{1}{N!}\left(\frac{V}{\lambda^3}\right)^N. \tag{2.11}$$

and whereas (2.8) and (2.10) are unaffected, the entropy becomes*

$$S = \frac{3}{2}Nk + Nk\ln\left(\frac{V}{N}\frac{1}{\lambda^3}\right). \tag{2.12}$$

Notice that now S/N no longer depends upon V, but upon the ratio V/N. The entropy has become an extensive quantity (i.e. proportional to N). Equation (2.12) is known as the *Sackur-Tetrode* equation.

It follows from (2.12) that the entropy of mixing of two gases is

$$\nabla S = N_1 k \ln\left[\left(\frac{V}{N}\right)\left(\frac{N_1}{V_1}\right)\right] + N_2 k \ln\left[\left(\frac{V}{N}\right)\left(\frac{N_2}{V_2}\right)\right].$$

When the gases are identical, there is a unique specific volume and therefore the entropy of mixing vanishes. Thus, the Gibbs paradox is resolved. When the gases are not identical, then of course $\Delta S > 0$, in agreement with the previous result.

2.5 Theorem of Equipartition of Energy

Suppose that the Hamiltonian of a system of particles is a quadratic function of the coordinates and of the momenta

$$H = \sum_i (a_i p_i^2 + b_i q_i^2). \tag{2.13}$$

* We use Stirling's approximation

$$N! \sim \sqrt{(2\pi)}\, e^{-N} N^{N+\frac{1}{2}} \qquad \text{for} \quad N \gg 1$$

in the form

$$\ln N! \sim N \ln N$$

The average energy corresponding to the particular term $a_i p_i^2$ is

$$\frac{a_i \int_0^\infty p_i^2 \exp(-a_i p_i^2/kT)\, dp_i}{\int_0^\infty \exp(-a_i p_i^2/kT)\, dp_i}$$

$$= \frac{-a_i kT(\partial/\partial a_i) \int_0^\infty \exp(-a_i p_i^2/kT)\, dv_i}{\int_0^\infty \exp(-a_i p_i^2/kT)\, dp_i}$$

$$= -a_i kT(\partial/\partial a_i)\left[\ln \int_0^\infty \exp(-a_i p_i^2/kT)\, dp_i\right]$$

$$= -a_i kT \frac{\partial}{\partial a_i}\left[\ln\left(\frac{\pi kT}{a_i}\right)^{\frac{1}{2}}\right] = \frac{1}{2}kT.$$

The same result is obtained for the particular term $b_i q_i^2$.

In other words, each term in the Hamiltonian which depends quadratically on a momentum or on a coordinate, contributes a mean energy of $\frac{1}{2}kT$.

This result is known as the theorem of *equipartition of energy*.

As an application, consider a solid containing N non-interacting particles harmonically bound to centres of forces. The Hamiltonian of each atom is

$$H = \sum_{i=1}^{3} \frac{p_i^2}{2m} + \frac{m\omega^2}{2} \sum_{i=1}^{3} q_i^2,$$

where ω is the angular frequency of oscillation. The first term is the kinetic energy of the atom, and the second term its potential energy of vibration.

The above theorem immediately gives

$$\text{Energy/atom} = 6 \times \tfrac{1}{2}kT = 3kT,$$

and for N atoms

$$E = 3NkT.$$

The specific heat of the solid is

$$C_V = (\partial E/\partial T)_V = 3Nk. \tag{2.14}$$

which is the law of *Dulong–Petit*.

2.6 Perfect Boson and Fermion Gases

We shall first calculate the mean occupation numbers $\bar{n}_i$ for boson and fermion gases using the grand canonical distribution. We shall then show that the same results can be obtained by using the canonical distribution, although this will oblige us to introduce a mathematical technique for evaluating certain integrals.

Referring to equation (1.56),

$$Q = \sum_n \sum_i z^n \exp(-E_i^n/kT). \tag{2.15}$$

Notice that each microscopic state of a system of indistinguishable particles is specified by a certain distribution of the occupation numbers of the particles, n_i, which are subject to the constraint

$$\sum_i n_i = n. \tag{2.16}$$

Such a distribution will be denoted by the symbol $\{n_i\}$. The number of accessible states of a system will then be given by all possible distributions $\{n_i\}$. Thus (2.15) can be rewritten as

$$Q = \sum_{n=0}^{\infty} \sum_{\{n_i\}} z^{\sum_i n_i} \exp\left(-\frac{1}{kT} \sum_i n_i E_i\right)$$

$$= \sum_{n=0}^{\infty} \sum_{\{n_i\}} \prod_i z^{n_i} \exp\left(-\frac{n_i E_i}{kT}\right) \tag{2.17}$$

The series in (2.17) contains only positive terms. Assuming that this series converges, one can, using a well-known theorem on absolutely convergent series, rearrange the terms and obtain

$$Q = \sum_{n_i} \prod_i z^{n_i} \exp(-n_i E_i/kT), \tag{2.18}$$

where now the sum is over all possible n_i without any restriction of the type (2.16). The reader can convince him elf of the equivalence of (2.17) and (2.18) by observing that each term in (2.17) appears in (2.18) and *vice versa*.

For a fermion gas, the occupation numbers can only take on the two values 0 or 1 (see section 2.2). Therefore, from (2.18) one obtains for fermions

$$Q = \prod_i [1 + z \exp(-E_i/kT)] \qquad \text{(fermions)}. \tag{2.19}$$

For bosons, on the other hand, n_i can be any non-negative integer.

Thus

$$Q = \prod_i [1+z\exp(-E_i/kT)+z^2\exp(-2E_i|kT)+\ldots]$$

$$= \prod_i [1-z\exp(-E_i/kT)]^{-1} \qquad \text{(bosons).} \tag{2.20}$$

In order to arrive at (2.20) we have assumed that

$$z\exp(-E_i/kT) < 1.$$

We have, then

$$\frac{PV}{KT} = \ln Q = \pm\sum_i \ln[1 \pm \exp(-E_i/kT)] \qquad \begin{matrix}+ \text{ for fermions} \\ - \text{ for bosons}\end{matrix} \tag{2.21}$$

and, according to (1.62), the mean total number of particles is given by

$$N = z\frac{\partial \ln Q}{\partial z} = \sum_i \frac{1}{z^{-1}\exp(E_i/kT)\pm 1}. \tag{2.22}$$

For the mean occupation number in an eigenstate i which has an energy E_i we find

$$\bar{n}_i = -kT\frac{\partial \ln Q}{\partial E_i} = \exp \frac{1}{\exp\dfrac{(E_i-\mu)}{kT} \pm 1} \qquad \begin{matrix}+ \text{ for bosons,} \\ - \text{ for fermions.}\end{matrix} \tag{2.23}$$

The expression (2.23) with a minus sign is called a *Bose–Einstein* distribution; with the plus sign it is called a *Fermi–Dirac* distribution.

One can also calculate the mean occupation number for a Boltz–mann gas. Here the numbers n_i can again be any non-negative integer. However, one must remember that since Boltzmann particles are distinguishable, to a given distribution $\{n_i\}$ there will correspond

$$\frac{1}{n!}\left(\frac{n!}{\prod_i n_i!}\right) = \frac{1}{\prod_i n_i!} \tag{2.24}$$

states of the system, instead of just one state, as in the case of a quantal gas. This result can be derived by noting that the first n_1 particles can be chosen in $\binom{n}{n_1}$ ways, the second group of n_2 particles

* The binomial coefficient $\binom{x}{y}$ is equal to $\dfrac{x!}{y!(x-y)!}$.

in $\binom{n-n_1}{n_2}$ ways, etc. Thus there are

$$\binom{n}{n_1}\binom{n-n_1}{n_2}\cdots = \frac{n!}{n_1!\,n_2!\,n_3!\,\ldots}$$

possible ways to achieve a distribution $\{n_i\}$. In addition, this condition must be multiplied by a factor $1/n!$ in order to take into account the reduction of the phase space available to a Boltzmann gas (cf. section 2.4). This leads to (2.24).

The grand partition function is then, taking (2.24) into account

$$Q = \prod_i \sum_{n_i} \frac{1}{n_i!} [z \exp(-E_i/kT)]^{n_i}$$

$$= \prod_i \exp[z \exp(-E_i/kT)].$$

Thus

$$\frac{PV}{kT} = \ln Q = \sum_i z \exp(-E_i/kT),$$

$$N = z\frac{\partial}{\partial z} \ln Q = \sum_i z \exp(-E_i/kT).$$

From the last two equations we obtain the equation of state of a perfect classical gas. The mean occupation numbers are given by

$$\bar{n}_i = -kT\frac{\partial \ln Q}{\partial E_i} = z \exp(-E_i/kT), \tag{2.25}$$

which is the well-known Boltzmann distribution.

Note that for bosons, the distribution of occupation numbers contains a singularity where the denominator vanishes. This fact has important physical consequences which we shall discuss in section 2.11.

2.7 The Darwin–Fowler Method

In order to show that the canonical distribution leads to the same results as the grand canonical distribution, we shall derive once more the results of the preceding section starting now from the partition function

$$q = \sum_{\{n_i\}} \exp[-\sum_i (n_i E_i)/kT], \qquad \sum_i n_i = N.$$

The restriction $\sum_i n_i = N$, with no subsequent summation over N, makes a direct evaluation of q difficult. We shall proceed in a less direct way.

Let us define a function $g(z)$ of the complex variable z (for the time being, z should not be confused with the activity).

$$g(z) = \sum_{n_i} z^{\sum_i n_i} \exp(-\sum_i n_i E_i/kT).$$

The function $g(z)$ is called a generating function. It is evident that the partition function q can be obtained from $g(z)$ by picking out of $g(z)$ the coefficient of z^N. Now, if we evaluate the sum in this equation in the same way as we evaluated the sums in (2.19) and (2.20), we find

$$g(z) = \prod_i \sum_{n_i} z^{n_i} \exp(-n_i E_i/kT) = \prod_i [1 \pm z \exp(-E_i/kT)]^{\pm 1}$$

so that $g(z)$ can be written as a Taylor series in z,

$$g(z) = \sum_{k=0}^{\infty} \frac{g^{(k)}(0)}{k!} z^k.$$

The coefficient of z^N is given by the well-known formula

$$\frac{g^{(N)}(0)}{N!} = \frac{1}{2\pi i} \int_C \frac{g(z)}{z^{N+1}} dz,$$

where C is a contour in the complex z-plane surrounding the origin. Thus

$$q = \frac{1}{2\pi i} \int_C \frac{g(z)}{z^{N+1}} dz, \tag{2.26}$$

and there remains to evaluate this integral. We shall do it using a method called the *method of steepest descent* (or the saddle point method), discovered by P. Debye.*

* A similar but more detailed account of this method can be found in *Mathematics for Physicists* by P. Dennery and A. Krzywicki, Harper and Row, New York, 1967.

Consider the integral

$$I(N) = \frac{1}{2\pi i} \int_C e^{N f(z)}\, dz, \tag{2.27}$$

where $f(z)$ is an analytic function of z, N a very large, real number, and C a contour in the z-plane.

The principal contribution to the integral comes from the integration region where $\mathrm{Re}\, f(z)$ has a maximum. This is true provided the oscillations of the phase $i\mathrm{Im}\, f(z)$ do not cancel this contribution. Debye's idea is to deform the contour C into a contour C_0 which is such that on C_0 the following conditions are satisfied:

(i) $\mathrm{Im}\, f(z)$ is constant along C_0,

(ii) C_0 goes through a point z_0 at which

$$\left.\frac{df}{dz}\right|_{z=z_0} = 0,$$

(iii) $\mathrm{Re}\, f(z)$ has a relative maximum at $z = z_0$.

Since an analytic function can have neither an absolute minimum nor an absolute maximum at a regular point z_0, condition (ii) defines a 'saddle point' of $f(z)$ which will also be a saddle point of $\mathrm{Re}\, f(z)$ and of $\mathrm{Im}\, f(z)$. The reason such a point is called a saddle point, is that the surface S which represents $\mathrm{Re}\, f(z)$ as a function of $\mathrm{Re}\, z$ and $\mathrm{Im}\, z$, resembles a horse's saddle in the neighbourhood of the point (see Fig. 1).

The following properties of the surface S, which are evident from Fig. 1, can be proved.

(a) The projections on the z-plane of the curves on S where $\mathrm{Re}\, f(z_0) = \mathrm{Re}\, f(z)$, divide the z-plane in four sectors labelled I, II, III, and IV.

(b) In each of these sectors there exists only one curve which is the projection of a curve on S where $\mathrm{Im}\, f(z) = \mathrm{Im}\, f(z_0)$.

(c) Only in two of the four sectors, can condition (iii) be satisfied for C_0.

In Fig. 1, the two sectors II and IV contain the projections of that part of the surface S where

$$\mathrm{Re}\, f(z) \leqslant \mathrm{Re}\, f(z_0). \tag{2.28}$$

Let us put

$$f(z) = f(z_0) - \tau^2,$$

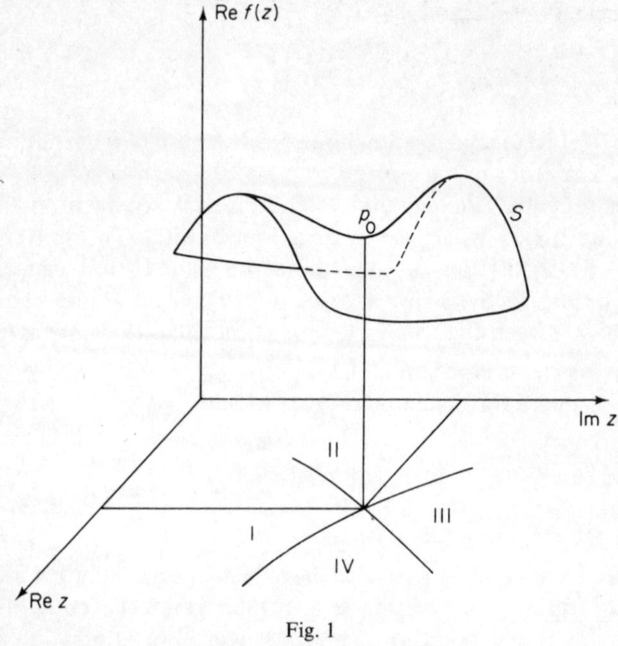

Fig. 1

where, on account of (i), τ must be real along C_0. The negative sign in front of τ^2 is chosen in such a way that (2.28) is satisfied along C_0. In an immediate neighbourhood of z_0, one has the Taylor expansion

$$\tau^2 = -\frac{1}{2}(z-z_0)^2 \left.\frac{d^2f}{dz^2}\right|_{z=z_0} + \ldots,$$

whence, in first approximation,

$$(z-z_0) = \frac{\tau e^{i\theta}\sqrt{2}}{|f''(z_0)|^{\frac{1}{2}}},$$

where θ is the phase angle of $(z-z_0)$. Thus (2.27) becomes

$$I(N) = \frac{e^{Nf(z_0)}}{2\pi i}\int\limits_{C_0} e^{-N\tau^2}\frac{dz(\tau)}{d\tau}\,d\tau$$

$$= \frac{1}{2\pi i}\frac{e^{Nf(z_0)}\,e^{i\theta}\sqrt{2}}{|f''(z_0)|^{\frac{1}{2}}}\int\limits_{C_0} e^{-N\tau^2}\,d\tau.$$

Since we are supposing that the most important contribution

to the integral comes from an immediate neighbourhood of z_0 ($\tau \ll 1$), one can replace the integral along C_0 by an integral along $(-\infty, \infty)$ which can then be easily evaluated. One obtains(*)

$$I(N) = \frac{1}{2\pi i} \frac{e^{Nf(z_0)}\, e^{i\theta} \sqrt{2}}{|f''(z_0)|^{\frac{1}{2}}} \sqrt{\frac{\pi}{N}}.$$

For $N \gg 1$, one has

$$\ln I(N) \cong N\, f(z_0).$$

Let us write (2.26) as follows

$$q = \frac{1}{2\pi i} \int_{C} e^{Nf(z)}\, dz,$$

where

$$\begin{aligned} N\, f(z) &= \ln g(z) - (N+1) \ln z \\ &\cong \ln g(z) - N \ln z. \end{aligned}$$

Then

$$\ln q \cong \ln I(N) \cong N\, f(z_0).$$

The point z_0, found from condition (ii), is the solution of the equation

$$N = \sum_i \frac{1}{z_0^{-1}\, e^{E_i/kT} \pm 1}.$$

The mean occupation number is given by

$$\bar{n}_i = -kT \frac{\partial \ln q}{\partial E_i} = \frac{1}{z_0^{-1}\, e^{E_i/kT} \pm 1}. \tag{2.29}$$

By identifying z_0 with the activity, one recovers the results of the preceding section.

2.8 The Boltzmann Limit of Boson and Fermion Gases

We wish to consider in this section the Bose–Einstein and Fermi–Dirac distributions in the Boltzmann limit, i.e. in the limit when the symmetry properties of the wave functions can be neglected.

Consider either the wave function (2.1) which describes a boson gas, or the wave function (2.2) which describes a fermion gas. The

* It is possible to show that $\theta = \frac{1}{2}\pi$ and hence q is real.

probability density for each gas will be given by the square of the modulus of its respective wave function. However, if the gas is rarefied, each particle will be localizable within a well-determined region which will be spatially separated from regions containing other particles. Thus the cross terms in the probability density, which represent the probabilities that one such region overlaps with another, will be negligible. It follows that for boson gases as well as for fermion gases,

$$|\psi_{\text{S.}}|^2 = |\psi_{\text{A.S.}}|^2 = \sum_P P|\psi(1, 2, \ldots, N)|^2.$$

This expression is simply the probability density of a system composed of N distinguishable particles which, however, we do not wish to distinguish experimentally, since a sum over all possible permutations is carried out. In other words it is simply the probability density for a system of N Boltzmann particles.

If the gas is rarefied, then we have

$$n_i \ll 1.$$

This condition will be satisfied at high temperatures because then the agitation of the particles of the gas will favor their separation.

In order to have an idea of the magnitude of the activity z at high temperatures, we shall calculate the chemical potential μ in the Boltzmann limit. From (1.53),

$$\frac{\mu}{\mathrm{T}} = -\left(\frac{\partial \mathrm{S}}{\partial \mathrm{N}}\right)_{V,E}.$$

Expressing the entropy given by (2.12) as a function of the energy given by (2.10), we get

$$\frac{\mu}{T} = -k\ln\left(\frac{V}{N}\frac{1}{\lambda^3}\right),$$

where λ is defined in (2.4). The chemical potential of a gas is negative in the Boltzmann limit, and its activity

$$z = e^{\mu/kT} = (N/V)\lambda^3$$

is equal to ratio of the particle 'volume' λ^3 to the specific volume V/N of the gas.

For a rarefied gas

$$\lambda^3 \ll (V/N).$$

Thus $z \ll 1$.

This condition allows one to make expansions in powers of z. From (2.22),

$$N = \sum_i \frac{z \exp(-E_i/kT)}{1 \pm z \exp(-E_i/kT)}$$

$$= \sum_{l=1} \sum_i (\pm)^{l+1} z^l \exp(-lE_i/kT).$$

As the thermodynamic limit is reached, we can replace the sum over the index i by an integral using (1.69) where, we recall,

$$\delta = \begin{cases} 1 & \text{for a quantal gas,} \\ 1/N! & \text{for a Boltzmann gas.} \end{cases} \tag{2.30}$$

Since

$$\int_0^\infty p^2 e^{-\alpha p^2} dp = \frac{1}{4} \frac{\sqrt{\pi}}{\alpha^{\frac{3}{2}}},$$

we find

$$\rho = \frac{N}{V} = \frac{1}{\lambda^3} \sum_{l=1}^{\infty} (\pm)^{l+1} \frac{z^l}{l^{\frac{3}{2}}}, \quad \begin{array}{l} - \text{ for fermions,} \\ + \text{ for bosons.} \end{array} \tag{2.31}$$

Similarly, we obtain from (2.21):

$$\frac{P}{kT} = \frac{1}{\lambda^3} \sum_{l=1}^{\infty} (\mp)^{l+1} \frac{z^l}{l^{\frac{5}{2}}}, \quad \begin{array}{l} - \text{ for fermions,} \\ + \text{ for bosons.} \end{array} \tag{2.32}$$

These two equations enable p/kT to be expanded, by successive eliminations of z, in powers of the density ρ. Such an expansion is called a *virial expansion*.

By retaining in (2.31) and (2.32) only the linear term in z,

$$\rho = \frac{z}{\lambda^3} + \ldots,$$

$$\frac{P}{kT} = \frac{2}{\lambda^3} + \ldots,$$

we immediately arrive at the perfect gas law

$$\frac{P}{kT} = \rho = \frac{N}{V}.$$

The result shows that at sufficiently high temperatures, all perfect gases reach their classical limit.

2.9 Equation of State of Perfect Gases

There exists a relation between the energy and the pressure of a gas which holds generally for all non-relativistic perfect gases.

Consider the expression (2.21). In the thermodynamic limit, one can replace the sum by an integral (see 1.69 and 2.30)

$$\frac{PV}{kT} = \pm\delta V \int_0^\infty \ln[1 \pm z \exp(-p^2/2mkT)] \frac{4\pi p^2\, dp}{h^3}.$$

By a partial integration, one obtains (since the integrated term vanishes)

$$\frac{P}{kT} = \frac{4\pi\delta}{3mkTh^3} \int_0^\infty \cdot \frac{z \exp(-p^2/2mkT)}{1 \pm z \exp(-p^2/2mkT)}\, p^4\, dp.$$

But the average energy of a gas is

$$E = V\delta \int_0^\infty \frac{p^2}{2m} \frac{1}{z^{-1} \exp(p^2/2mkT) \pm 1} \frac{4\pi p^2\, dp}{h^3}.$$

By comparison we find the general equation

$$P = \frac{2}{3}\frac{E}{V}, \tag{2.33}$$

which reduces to (2.8) in the particular case of a classical gas for which the energy is given by (2.10).

The reader can verify that for a relativistic gas where the energy of each particle depends linearly on the momentum,

$$\varepsilon = cp,$$

the expression (2.33) is replaced by

$$P = \frac{1}{3}\frac{E}{V}, \tag{2.34}$$

2.10 Degenerate Gases

A gas is said to be *degenerate* if the effects of symmetrization are important. A gas will therefore be degenerate at low temperatures. The properties of a degenerate fermion gas are very different from those of a degenerate boson gas. For this reason we shall consider them separately. First, we state an important theorem that applies to both bosons and fermions.

The Theorem of Nernst The entropy of a supersystem A + B in thermodynamic equilibrium is,

$$S_{AB} = -k \sum_{i,j} p_{i_A, j_B} \ln p_{i_A, j_B},$$

and the probabilities are given by the microcanonical distribution (1.15). Thus

$$S_{AB} = k \ln \Omega.$$

To the extent that the sub-systems A and B can be considered as isolated one from the other or, more precisely, if there exists a quasi-stationary distribution of the energy and of the number of particles between A and B, we should also be able to write for system A,

$$S_A = k \ln \Omega_A,$$

where Ω_A is the number of accessible states of A which have the same energy and the same number of particles.

At absolute zero temperature, Ω_A will be equal to unity when there is only one state of the system having the lowest possible energy. In that case the entropy will vanish.

But, even if at absolute zero, $\Omega \neq 1$, this quantity will nevertheless not be strongly N-dependent, so that the entropy per molecule of the system will very nearly vanish. In the limit $N \to \infty$, one will have

$$\lim_{N \to \infty} \frac{S_A}{N} = \lim_{N \to \infty} \frac{k}{N} \ln \Omega_A = 0.$$

This is an expression of the theorem of Nernst, often known as the third law of thermodynamics.

The argument just given, which leads to Nernst's theorem, applies only to a quantal system. In fact, in classical statistical mechanics, entropy is defined up to an arbitrary additive constant only.

The completely Degenerate Fermion Gas A gas at $T = 0$ is said to be completely degenerate. A fermion gas at absolute zero temperature has the remarkable property that its total kinetic energy is not zero. This is an immediate consequence of the Pauli exclusion principle; for, since each quantum state can contain only one fermion, it follows that the fermions in the gas will occupy all possible quantum states whose energies lie between zero and a quantity E_F, called the *Fermi energy*, whose value will depend upon the number of fermions in the gas.

The Fermi–Dirac distribution at absolute zero will thus be a step-function

$$\bar{n}_i = \begin{cases} 1 & \text{if } E_i \leqslant E_F, \\ 0 & \text{if } E_i > E_F. \end{cases} \tag{2.35}$$

Comparing (2,35) with the Fermi–Dirac distribution for finite temperatures

$$\bar{n}_i = \frac{1}{\exp[(E_i - \mu)/kT] + 1},$$

and denoting by μ_F the chemical potential at $T = 0$, one sees that $\bar{n}_i \to 0$ when $T \to 0$ only if $E_i > \mu_F$. Thus, the Fermi energy E_F coincides with the chemical potential at $T = 0$,

$$E_F = \mu_F.$$

Putting $T = 0$ in the thermodynamic relation (1.68) and using (2.33), we get

$$\frac{E}{N} = \frac{3}{5}\mu_F,$$

$$P = \frac{2}{5}\frac{N}{V}\mu_F,$$

which shows that even at absolute zero, the pressure of a fermion gas is not zero.

The Fermi energy μ_F itself depends upon the density of the gas, as a simple calculation shows. Using (2.35), the total number of particles at $T = 0$ is

$$N = g_0 V \int_0^{p_F} \frac{4\pi p^2\, dp}{h^3}$$

$$= \frac{4\pi g_0}{h^3}(2m\mu_F)^{\frac{3}{2}} \qquad \left(\mu_F = \frac{p_F^2}{2m}\right),$$

where the factor

$$g_0 = (2S+1)$$

is the degeneracy of states of a particle with spin S. Thus

$$\mu_F = \frac{1}{2m}\left(\frac{3h^3 N}{4\pi g_0 V}\right)^{\frac{2}{3}}.$$

Defining the *Fermi temperature*

$$T_F = \mu_F/k, \tag{2.36}$$

it is seen that 'low temperature' means

$$T \ll T_F,$$

and that therefore this notion depends strongly upon the density of the gas considered, and upon the masses of its constituent particles.

We shall now generalize some of the preceding results to the case when the particles move with relativistic velocities, for it is in this case that the theory of completely degenerate Fermi gases finds some of its most interesting applications. This will be seen in the following section.

The pressure of a gas, when the particles move isotropically with velocity v is given by the expression

$$P = \frac{1}{3V}\int_0^\infty n_i(p)\, pv(p)\left(\frac{4\pi g_0 V}{h^3}\right)p^2\, dp.$$

Using (2.35) one finds for a completely degenerate gas:

$$P = \frac{4\pi g_0}{3h^3}\int_0^{p_F} p^3 v(p)\, dp,$$

where p is the momentum corresponding to the energy E.
In general, the velocity $v(p)$ will be given by the relativistic expression

$$v(p) = cp/\sqrt{(m^2c^2+p^2)},$$

so that

$$P = \frac{4\pi g_0 c}{3h^3}\int_0^{pF}\frac{p^4}{\sqrt{(m^2c^2+p^2)}}\, dp.$$

Putting $p/mc = \sinh\theta$,

$$P = \frac{4\pi g_0 m^4 c^5}{3h^3}\int_0^{\theta_F}\sinh^4\theta\, d\theta.$$

The integral can be found in tables. Putting

$$x = p_F/mc, \qquad \text{i.e. } \sinh\theta_F = x,$$

we get

$$P = \frac{\pi g_0 m^4 c^5}{6h^3} f(x), \tag{2.37}$$

where

$$f(x) = x(1+x^2)^{\frac{1}{2}}(2x^2-3)+3\sinh^{-1}x. \tag{2.38}$$

Equations (2.37) and (2.38), which may be rewritten in terms of the parameter x as

$$\rho = \frac{4\pi g_0 m^3 c^3}{3h^3} x^3, \tag{2.39}$$

form the basis for the study of the completely degenerate gas. We shall consider these equations in the two important limiting cases: $x \ll 1$ and $x \gg 1$.

(i) when $x \ll 1$ (non-relativistic case) the function $f(x)$ can be expanded as follows:

$$f(x) \sim \frac{8}{5}x^5 - \frac{4}{7}x^7 + \frac{1}{3}x^9 + \ldots.$$

From (2.37) and (2.39) one readily obtains

$$P = \frac{h^2}{5m}\left(\frac{3}{4\pi g_0}\right)^{\frac{2}{3}} \rho^{\frac{5}{3}}. \tag{2.40}$$

The internal energy of the gas in the non-relativistic limit is given by

$$\frac{E}{V} = \frac{4\pi g_0}{h^3}\int_0^{p_F} \frac{p^2}{2m} p^2\, dp$$

$$= \frac{2\pi g_0}{5mh^3} p_F^5 = \frac{3h^2}{10m}\left(\frac{3}{4\pi g_0}\right)^{\frac{2}{3}} \rho^{\frac{5}{3}}. \tag{2.41}$$

Combining (2.40) and (2.41), we recover once again equation (2.33).

(ii) when $x \gg 1$ (relativistic case) we have

$$f(x) \sim 2x^4 - 3x^2 + \ldots.$$

Thus

$$P = \frac{hc}{4}\left(\frac{3}{4\pi g_0}\right)^{\frac{1}{3}} \rho^{\frac{4}{3}}. \tag{2.42}$$

The internal energy is now given by

$$\frac{E}{V} = \frac{4\pi g_0}{h^3} \int_0^{p_F} (cp)\, p^2\, dp$$

$$= \frac{\pi g_0 c}{h^3} p_F^4 = \frac{3hc}{4}\left(\frac{3}{4\pi g_0}\right)^{\frac{1}{3}} \rho^{\frac{4}{3}}. \tag{2.43}$$

Equations (2.42) and (2.43) yield the general relation (2.34) for a relativistic gas.

The expressions (2.40) and (2.42) show that, even at $T = 0$, the pressure of a fermion gas does not vanish. This again is a consequence of the Pauli principle.

White Dwarfs In this section we illustrate the use of the degenerate electron gas theory with an application borrowed from stellar dynamics.

White dwarfs have three characteristic features. Compared to other stars of the same mass they are much fainter, they have a much smaller diameter, and they are very dense. It has been shown by R. H. Fowler and by S. Chandrasekhar that such stellar configurations can be well described by treating the electrons within the stellar matter as a completely degenerate relativistic Fermi gas. The reason is that the very high electron density yields on the one hand, a Fermi temperature which is much greater than the temperature of the white dwarf, and on the other hand causes the electrons to attain relativistic energies.

In this section we shall show that the completely degenerate Fermi gas theory leads to the prediction that the mass of a white dwarf cannot, without collapse, exceed a certain limit, which is of the order of the mass of the sun. Consider a spherical distribution of matter of density $\omega(r)$ and let $V(r)$ represent the attractive gravitational potential at a distance r from the centre of the distribution. Then $V(r)$ obeys Poisson's equation which, for a spherically symmetric distribution reads

$$\frac{1}{r^2}\frac{d}{dr}\left(r^2 \frac{dV}{dr}\right) = 4\pi G\omega(r),$$

where G is the gravitational constant. A completely degenerate Fermi gas of electrons, as we have seen, exerts an outward pressure, even at $T = 0$. If dP is the difference in that pressure across a shell of thickness dr, we have, in an equilibrium configuration

$$-dP = \frac{dV}{dr}\omega\, dr.$$

Combining

$$\frac{1}{r^2}\frac{d}{dr}\left(\frac{r^2}{\omega}\frac{dP}{dr}\right) = -4\pi G\omega. \tag{2.44}$$

From this equation we can deduce some interesting properties. Putting

$$\omega(r) = \mu\rho(r),$$

where μ is the molecular weight of the stellar matter, and after using (2.37), (2.38), and (2.39), (2.44) becomes after some elementary calculations

$$\frac{1}{r^2}\frac{d}{dr}\left(r^2\frac{d(x^2+1)^{\frac{1}{2}}}{dr}\right) = -\Omega x^3, \tag{2.45}$$

where $\Omega = 32\pi^2 m^2 c\mu^2 G/3h^3$. Eq. (2.45) can be conveniently transformed. Let

$$y^2 = x^2+1,$$

and let x_0 be the density of matter at the centre of the stellar distribution. The corresponding value of y will be denoted by y_0. Finally put

$$y = y_0\,\theta(r), \tag{2.46}$$

$$r = \frac{1}{y_0\sqrt{\Omega}}\eta,$$

In terms of θ and η, (2.45) becomes

$$\frac{1}{\eta^2}\frac{d}{d\eta}\left(\eta^2\frac{d\theta}{d\eta}\right) = -\left(\theta^2-\frac{1}{y_0^2}\right)^{\frac{3}{2}}. \tag{2.47}$$

We shall be interested in the special case of very high central densities x_0. Then $y_0 \to \infty$ and (2.47) reduces to

$$\frac{1}{\eta^2}\frac{d}{d\eta}\left(\eta^2\frac{d\theta}{d\eta}\right) = -\theta^3. \tag{2.48}$$

On account of (2.46) and of regularity conditions at the origin, the boundary conditions associated with (2.48) are

$$\theta(0) = 1, \qquad \left(\frac{d\theta}{d\eta}\right)_{\eta=0} = 0. \tag{2.49}$$

Equation (2.48) is called a Lane–Emden differential equation and its solution which satisfies the boundary conditions (2.49) is called a Lane–Emden function of order 3 and denoted by θ_3.*

The mass of stellar matter contained within a sphere of radius R is

$$M = 4\pi \int_0^R \omega r^2 \, dr = \frac{4\pi\mu\alpha}{\Omega^{\frac{3}{2}}} \int_0^{\eta_0} \theta^3 \eta^2 \, d\eta,$$

where $\alpha \equiv 8\pi m^3 c^3/3h^3$.
Using (2.48) which holds when x_0 is large, this becomes

$$M = \frac{-4\pi\mu\alpha}{\Omega^{\frac{3}{2}}} \left(\eta^2 \frac{d\theta_3}{d\eta} \right)_{\eta=\eta_0}.$$

Of course, η_0 should correspond to the radius where the density of matter or, (what is the same when x_0 is large) where θ_3 vanishes. From tables of Lane–Emden functions, one finds†

$$\eta_0 = 6{\cdot}897,$$
$$\left(\eta^2 \frac{d\theta_3}{d\eta} \right)_{\eta_0} = -2{\cdot}018.$$

Thus the total mass of the stellar matter is

$$M = \frac{4\pi\mu\alpha}{\Omega^{\frac{3}{2}}} 2{\cdot}018. \qquad (2.50)$$

Expressed in terms of the mass of the sun, $\odot = 1{\cdot}985 \times 10^{33}$ g, (2.50) yields

$$M = \frac{5{\cdot}7}{\mu^2} \odot.$$

As an example, the mean molecular weight of Sirius B being $\mu = 1{\cdot}3$, its mass cannot exceed a value of the order of $3\odot$.

The Degenerate Fermi Gas We shall now consider a perfect fermion gas at low but finite temperatures. More precisely, we shall assume that

$$\mu/kT = \ln z \gg 1. \qquad (2.51)$$

In order to simplify the notation, we introduce the function

$$f\left(\frac{E-\mu}{K} \right) \equiv \frac{1}{e^{(E-\mu)/K} + 1}, \qquad (K = kT).$$

* The order is the power to which the function on the RHS of (2.48) is raised.
† See reference 2 Table 4, p. 96.

We wish to calculate average values of the type

$$\overline{h(\mu)} = \int_0^\infty h(E) f(E - \mu/K) dE, \tag{2.52}$$

where $h(E)$ is some function of the energy.* Putting

$$H(E) \equiv \int_0^E h(E')\, dE', \qquad y = (E - \mu)/k,$$

and integrating (2.52) by parts, we get

$$\overline{h(\mu)} = \int_{-\mu/K}^{\infty} H(Ky + \mu) \frac{df}{dy}\, dy.$$

But the condition (2.51) allows one to replace the lower limit of the integral above by $-\infty$. Therefore

$$\overline{h(\mu)} \cong + \int_{-\infty}^{+\infty} H(Ky + \mu) \frac{df}{dy}\, dy. \tag{2.53}$$

Now the expansion

$$H(Ky + \mu) = H(\mu) + Ky \frac{\partial H(\mu)}{\partial \mu} + \frac{1}{2}(Ky)^2 \frac{\partial^2}{\partial \mu^2} H(\mu) + \ldots.$$

allows one to write (at least formally)

$$H(Ky + \mu) = \exp(Ky\, \partial/\partial u)\, H(\mu),$$

as can be seen by expanding the exponential above. Hence, (2.53) can be written

$$\overline{h(\mu)} = \left[\int_{-\infty}^{\infty} \exp\left(Ky \frac{\partial}{\partial u} \right) \frac{e^y}{(e^y + 1)^2}\, dy \right] H(\mu),$$

or, putting $e^y = \xi$,

$$\overline{h(\mu)} = \left[\int_0^\infty \frac{\xi^{K\, \partial/\partial \mu}}{(\xi + 1)^2}\, d\xi \right] H(\mu).$$

The integral above can be evaluated using the method of contour integration, with the result†

$$\overline{\mathrm{h}(\mu)} = \left[\pi K \frac{\partial}{\partial \mu} \csc \pi K \frac{\partial}{\partial \mu} \right] H(\mu)$$

* Here we shall follow a method due to R. Blankenbecler. See C. Kittel, *Elementary Statistical Physics*, New York: J. Wiley; London: Chapman and Hall (1961).

† See reference 4.

$$= \left[1+\frac{\pi^2}{6}K^2\frac{\partial^2}{\partial\mu^2}+\frac{7\pi^4}{360}K^4\frac{\partial^4}{\partial\mu^4}+\ldots\right]H(\mu).$$

We can use this to calculate averages. The average energy of a gas is

$$E = Vg_0\int E\,f\left(\frac{E-\mu}{K}\right)\frac{d^3p}{h^3}$$

$$= \frac{(2\pi V)g_0\,(2m)^{\frac{3}{2}}}{h^3}\int_0^\infty E^{\frac{3}{2}}f\left(\frac{E-\mu}{K}\right)dE.$$

Thus in this case we should set

$$H(\mu) = \frac{2}{5}\frac{(2\pi V)g_0\,(2m)^{\frac{3}{2}}}{h^3}\mu^{\frac{5}{2}},$$

and E can be expanded as

$$E = \frac{2}{5}\frac{(2\pi V)g_0\,(2m)^{\frac{3}{2}}}{h^3}\mu^{\frac{5}{2}}\left[1+\frac{5\pi^2}{8}\left(\frac{kT}{\mu}\right)^2+\ldots\right].$$

Similarly, the number of particles is given by

$$N = \frac{(2\pi V)g_0\,(2m)^{\frac{3}{2}}}{h^3}\int_0^\infty f\left(\frac{E-\mu}{K}\right)E^{\frac{1}{2}}\,dE$$

$$= \frac{2}{3}\left(\frac{2\pi V g_0}{h^3}\right)(2m)^{\frac{3}{2}}\mu^{\frac{3}{2}}\left[1+\frac{\pi^2}{8}\left(\frac{kT}{\mu}\right)^2+\ldots\right]. \tag{2.54}$$

Expanding the ratio E/N in powers of (kT/μ) gives

$$\frac{E}{N} = \frac{3}{5}\mu\left[1+\frac{\pi^2}{2}\left(\frac{kT}{\mu}\right)^2+\ldots\right]. \tag{2.55}$$

At $T = 0$, from (2.54) we get

$$N = \frac{2}{3}\left(\frac{2\pi V g_0}{h^3}\right)(2m)^{\frac{3}{2}}\mu_F^{\frac{3}{2}}. \tag{2.56}$$

Combining (2.54) and (2.56)

$$\mu_F^{\frac{3}{2}} = \mu^{\frac{3}{2}}\left[1+\frac{\pi^2}{8}\left(\frac{kT}{\mu}\right)^2+\ldots\right].$$

This series can be inverted to give

$$\mu = \mu_F\left[1-\frac{\pi^2}{12}\left(\frac{kT}{\mu_F}\right)^2+\ldots\right].$$

Substituting in (2.55), we have

$$E = \frac{3}{5} N \mu_F \left[1 + \frac{5\pi^2}{12} \left(\frac{kT}{\mu_F} \right)^2 + \ldots \right], \tag{2.57}$$

which leads to the result already obtained for a completely degenerate gas at $T = 0$.

The specific heat at constant volume

$$C_V = \left(\frac{\partial E}{\partial T} \right)_V = \frac{\pi^2}{2} N \frac{k^2}{\mu_F} T \tag{2.58}$$

is linear with respect to T at low temperatures. We recall that it was independent of T at high temperatures.

Finally, from (2.33) and (2.57) we get

$$P = \frac{2}{5} \rho \mu_F \left[1 + \frac{5\pi^2}{12} \left(\frac{T}{T_F} \right)^2 + \ldots \right]. \tag{2.59}$$

2.11 Bose–Einstein Condensation

We will now consider the boson gas and some of its peculiar features.

The mean occupation numbers for a boson gas are given by

$$\bar{n}_i(E) = \frac{1}{z^{-1} \exp(E_i/kT) - 1}. \tag{2.60}$$

Since the ground state has zero energy, one must have $z^{-1} \geqslant 1$ in order for the occupation number to remain a non-negative quantity. In general, therefore,

$$0 \leqslant z \leqslant 1,$$

so that, for a boson gas, the chemical potential is non-negative at all temperatures,

$$\mu \leqslant 0. \tag{2.61}$$

For $z < 1$, we had (*cf.* 2.31, 2.32)

$$\frac{P}{kT} = \frac{1}{\lambda^3} \sum_{l=1}^{\infty} \frac{z^l}{l^{\frac{5}{2}}}, \tag{2.62}$$

$$\rho = \frac{1}{\lambda^3} \sum_{l=1}^{\infty} \frac{z^l}{l^{\frac{3}{2}}}. \tag{2.63}$$

The series in (2.62) and (2.63) converge only if $z \leqslant 1$. When the convergence limit $z = 1$ is reached, (2.63) determines, for *a given temperature*, a critical density ρ_c defined by

$$\rho_c = \frac{1}{\lambda^3} \sum_{l=1}^{\infty} \frac{1}{l^{\frac{3}{2}}} = \frac{1}{\lambda^3} \rho(\tfrac{3}{2}) = \frac{1}{\lambda^3} 2{\cdot}612, \tag{2.64}$$

where $\zeta(x)$ is the Riemann zeta function.

When $\mu < 0$, the right side of (2.63) decreases as z decreases and one obtains densities that are smaller then ρ_c. To the condition (2.61), therefore, there correspond densities

$$\rho \leqslant \rho_c.$$

The situation is very different, however, when, at a given temperature, the density of the gas is greater than the critical density ρ_c. Then equation (2.63) is no longer valid for such densities since, as we have just seen, this equation can only describe densities that are smaller than the critical density.

The reason for the contradiction is the following. In order to arrive at (2.63), we replaced a sum by an integral which contained a weight, namely, the density of states at a given energy. In the thermodynamic limit, the energy levels became dense. However, for a boson gas, the ground state can contain many particles since there are no limitations on the number of particles that can belong to a single quantum state. In spite of this possibility, the ground state was given zero weight since the density of states is proportional to $V\,d^3p$, i.e. to $VE^{\frac{1}{2}}dE$, and therefore vanishes when $E = 0$.

In order to remove the contradiction let us add to N the population of the ground state $N(0)$

$$N = N(0) + \frac{V}{\lambda^3} \sum_{l=1}^{\infty} \frac{z^l}{l^{\frac{3}{2}}}, \tag{2.65}$$

where $N(0) = 1/(z^{-1}-1)$.

The second term in (2.65) is the average number of particles in the excited states

$$N_{\text{exc}} = \frac{V}{\lambda^3} \sum_{l=1}^{\infty} \frac{z^l}{l^{\frac{3}{2}}}.$$

If (2.65) is solved graphically for z, we find in the limit $V \to \infty$, the following simple result:

$$z = 1 \quad \text{when} \quad \rho > \rho_c. \tag{2.66}$$

Thus when $\rho > \rho_c$

$$N_{\text{exc}} = \frac{V}{\lambda^3} 2{\cdot}612.$$

Defining now, *for a given density*, a critical temperature T_c, by the relation

$$N = \frac{V}{\lambda_c^3} 2{\cdot}612,$$

we obtain for $T < T_c$

$$N(0) = N - N_{\text{exc}} = N\left[1 - \frac{N_{\text{exc}}}{N}\right]$$

$$= N\left[1 - \left(\frac{T}{T_c}\right)^{\frac{3}{2}}\right],$$

which shows that $N(0)$ can be very large, even when T is only slightly smaller than T. The relation also shows that at $T = 0$, all bosons are in the ground state. One also has

$$\frac{P}{kT} = \left\{\frac{1}{V}\ln(1-z)\right\} + \frac{1}{\lambda^3}\sum_{l=1}^{\infty}\frac{z^l}{l^{\frac{5}{2}}},$$

and, for $\rho > \rho_c$ we find, using (2.66),

$$\frac{P}{kT} = \lim_{z\to 1}\left\{\frac{1}{V}\ln(1-z)\right\} + \frac{1}{\lambda^3}\zeta(\tfrac{5}{2})$$

$$= \lim_{z\to 1}\left\{\frac{1}{V}\ln(1-z)\right\} + \frac{1}{\lambda^3}1{\cdot}341. \tag{2.67}$$

But, for large V,

$$\mu = 0(1/V),$$

Hence

$$1 - z = 1 - e^{-\mu/kT} \sim \frac{\mu}{kT} = 0\left(\frac{1}{V}\right),$$

which means that the first term in (2.67) vanishes as $V \to \infty$. For

$T < T_c$, then

$$P = \frac{kT}{\lambda^3} 1{\cdot}341$$

$$= 1{\cdot}341 \left(\frac{2\pi m}{h^2}\right)^{\frac{3}{2}} (kT)^{\frac{5}{2}}. \tag{2.68}$$

The specific heat per particle is given by the expression

$$c_V = \frac{1}{N}\left(\frac{\partial E}{\partial T}\right) = \frac{3}{2}\frac{1}{\rho}\frac{dP}{dT}, \tag{2.69}$$

i.e.

$$c_V = \begin{cases} \dfrac{15}{4}(1{\cdot}341)\left(\dfrac{2\pi m}{h^2}\right)^{\frac{3}{2}} \rho k^{\frac{5}{2}} T^{\frac{3}{2}}, & T < T_c, \\ \dfrac{3}{2}\dfrac{1}{\rho}\dfrac{d}{dT}\left[\dfrac{kT}{\lambda^3}\displaystyle\sum_{l=1}^{\infty}\dfrac{z^l}{l^{\frac{5}{2}}}\right], & T > T_c. \end{cases} \tag{2.70}$$

Hence, the specific heat has a discontinuous derivative at $T = T_c$. This property is characteristic of the occurrence of a phase transition. The phenomenon is called a Bose–Einstein condensation.

From (2.68) one sees that in the critical region, the pressure varies as $T^{\frac{5}{2}}$ and is independent of the volume. In other words, the condensed state does not contribute to the pressure. Consequently, the phase transition occurs without a corresponding change in the volume. Hence the condensed state occupies no volume.

The entropy, which can be calculated from the expression

$$S = \int_0^T \frac{c_V}{T'}\, dT',$$

is given by

$$S = \begin{cases} \dfrac{5}{2}(1{\cdot}341)\left(\dfrac{2\pi m}{h^2}\right)^{\frac{3}{2}} \rho h^{\frac{5}{2}} T^{\frac{3}{2}}, & T < T_c, \\ \dfrac{5}{2}\left(\dfrac{2\pi m}{h^2}\right)^{\frac{3}{2}} \rho h^{\frac{5}{2}} T^{\frac{3}{2}} \displaystyle\sum_{l=1}^{\infty}\dfrac{z^l}{l^{\frac{5}{2}}} - \ln z, & T > T_c. \end{cases} \tag{2.71}$$

The entropy vanishes at $T = 0$, in agreement with Nernst's theorem.

Black Body Radiation Photons have unit spin and are therefore bosons. These bosons can be considered as a gas, and in fact as a perfect gas, since it follows from electromagnetic theory that their interactions are completely negligible. If a closed cavity is maintained at a temperature T, it will emit and reabsorb photons. After a certain lapse of time, a situation will be established wherein the photons and the matter of which the cavity is composed will be in thermodynamic equilibrium. The electromagnetic radiation within the cavity is called *black body radiation.* In this section we shall describe the properties of this radiation, and in the following section we shall consider an application of extreme importance.

Photons differ from other bosons in that their total number is not conserved. From the definition of the Helmholtz free energy (1.47) and from (1.53) we have

$$dA = -P\,dV + \mu\,dN - S\,dT$$

where dN is not equal to zero for a photon gas. But it is well known that the function A is a minimum with respect to changes of state which occur at constant T and V, when the state is in thermodynamic equilibrium. Thus

$$\left(\frac{\partial A}{\partial N}\right)_{V,T} = \mu = 0$$

and the chemical potential of a photon gas vanishes.

Since the energy of a photon is given by $E = \hbar\omega = h\nu$ where ν is its frequency, one has for the energy distribution of photons in thermodynamic equilibrium

$$\bar{n}_\nu = \frac{1}{e^{h\nu/kT} - 1}$$

The energy density per frequency interval $d\nu$ is (taking into account the two possible polarization states of a photon)

$$\varepsilon\,d\nu = \frac{8\pi p^2}{h^3}\bar{n}_\nu\,h\nu\,dp = \frac{8\pi h\nu^3}{c^3}\bar{n}_\nu\,d\nu$$

Therefore

$$\varepsilon = \frac{8\pi h}{c^3}\frac{\nu^3}{e^{h\nu/kT} - 1}$$

which is *Planck's radiation law.*

When $h\nu/kT \ll 1$, one obtains the classical limit of Planck's law

$$\varepsilon \cong \frac{8\pi h\nu^3}{c^3}\cdot\left(\frac{kT}{h\nu}\right) \tag{2.72}$$

and when $h\nu/kT \gg 1$ one obtains

$$\varepsilon \cong \frac{8\pi h\nu^3}{c^3} e^{-h\nu/kT}. \tag{2.73}$$

(2.72) is the *Rayleigh–Jeans* law and (2.73) is *Wien's* law.

Wien's law displays the fact that for each temperature T there will correspond a wavelength $\lambda \doteq \lambda_0$ for which ε will be a maximum. This wavelength is given by

$$\lambda_0 = 2{\cdot}9 \times 10^{-3}/T \text{ m °K}.$$

This result is known as *Wien's displacement* law. The total energy density in the radiation field is given by

$$\frac{E}{V} = \int_0^\infty \varepsilon \, d\nu = \frac{8\pi}{c^3 h^3} \frac{V}{(hc)^3} (kT)^4 \int_0^\infty \frac{x^3 \, dx}{e^x - 1}.$$

The integral is a constant equal to $\pi^4/15$. Thus

$$\frac{E}{V} = \frac{4\sigma V}{c} T^4, \tag{2.74}$$

where

$$\sigma = \frac{\pi^2 k^4}{60\hbar^3 c^2}$$

is called the Stefan–Boltzmann constant. The above result is known as the *Stefan–Boltzmann* law.

The Einstein derivation of the Planck radiation law. The maser and the laser An atom can emit radiation if an electron makes a transition from a higher to a lower state $m \to n$. The transition can be either spontaneous, or induced by the presence of external radiation of energy density ρ per unit frequency range. The number of spontaneous radiative emissions per unit time is

$$N_m A_{mn}, \tag{2.75}$$

where N_m is the number of atoms in the state m and A_{mn} is a coefficient of proportionality. The number of induced emissions, contrary to the number of spontaneous ones, will be proportional to the external radiation ρ. Their number will be

$$N_m B_{mn} \rho. \tag{2.76}$$

The coefficients A_{mn} and B_{mn} are called *Einstein coefficients.*

Thus far we have considered transitions from the state m to the state n. There will also be transitions from the state n to the state m, but since the energy of m is higher than the energy of n, these transitions can only be induced by the external radiation, and the number of absorptions will be

$$N_n B_{nm} \rho.$$

The spontaneous radiation will be incoherent, but the induced radiation will have the same phase as the external radiation.

When thermodynamic equilibrium is reached,

$$N_m A_{mn} + N_m B_{mn} \rho = N_n B_{nm} \rho,$$

i.e.

$$\rho = \frac{N_m A_{mn}}{N_n B_{nm} - N_m B_{mn}}.$$

Taking for the energy distribution of the atoms a Boltzmann distribution

$$\frac{N_m}{N_n} = e^{-h\nu/kT},$$

where $h\nu$ is the energy difference between the levels m and n, we find

$$\rho = \frac{A_{mn}}{e^{h\nu/kT} B_{nm} - B_{mn}}.$$

But B_{nm}, which is a transition probability, and therefore a matrix element of a hermitean Hamiltonian, is symmetric:

$$B_{nm} = B_{mn}. \tag{2.77}$$

It follows that

$$\rho = \frac{A_{mn}/B_{mn}}{e^{h\nu/kT} - 1}. \tag{2.78}$$

For very large T,

$$\rho \cong \left(\frac{A_{mn}}{B_{mn}}\right) \frac{kT}{h\nu}.$$

Comparing with the Rayleigh–Jeans law (2.72), one finds

$$\frac{A_{mn}}{B_{nn}} = \frac{8\pi h\nu^3}{c^3}. \tag{2.79}$$

With the relations (2.78) and (2.79) we are led once again to the Planck radiation law, and at the same time we will have given this law a physical interpretation. This derivation was first given by Einstein.

As we have seen, the radiative transitions from a state m to a state

n are given by

$$(A_{mn}+\rho B_{mn})\,N_m\,.$$

and the inverse transitions by

$$B_{nm}\,\rho N_n$$

If the atoms are immersed in a radiation field of density ρ, this incident radiation will emerge from the atoms with an intensity

$$(N_m-N_n)\,\rho B_{nm}$$

which will be either reduced or amplified with respect to the incident radiation, according as

$$N_m < N_n$$

or

$$N_m > N_n \tag{2.80}$$

It is possible to achieve a situation where (2.80) is satisfied even though the energy of m is greater than the energy of n. This situation, which one calls a 'population inversion', is a non-equilibrium situation. It gives rise to a particularly interesting and important effect which we will now briefly describe.

Let electromagnetic radiation of intensity $I(\nu)$ and of frequency lying in the range between ν and $\nu+d\nu$, travel in the x-direction and traverse a slab of matter of thickness dx, in a time $dt = dx/v$, where v is the velocity of the radiation in the matter. Let there be $N_{n\nu}$ atoms in level n which can absorb radiation in the frequency interval $d\nu$ around ν, and $N_{m\nu}$ atoms in level m which can emit radiation in the same interval. Since, as we have seen, the induced radiation will have the same phase as the external radiation, the change in the intensity of the external radiation as it traverses the slab dx in the time dt will be

$$-dI(\nu)\,d\nu = h\nu B_{nm}\,(dN_{n\nu}-dN_{m\nu})\frac{I(\nu)\,dx}{v} \tag{2.81}$$

where we have made use of (2.77). On the other hand, it is a well known experimental fact that the intensity of electromagnetic radiation which traverses matter varies according to the law

$$I(\nu) = I_0\,e^{-\kappa_\nu x} \tag{2.82}$$

where κ_ν is the co-called absorption coefficient.

(2.81) and (2.82) (written in differential form) yield, after an integration over the width of the absorption line

$$\int \kappa_\nu \, d\nu = \frac{h\bar{\nu}}{v} B_{nm} (N_n - N_m) \tag{2.83}$$

where $\bar{\nu}$ corresponds to the centre of the line.

Thus when the condition (2.80) of population inversion is realized, the integrated absorption coefficient of the incident radiation is negative. This means that the incident radiation is amplified as it emerges from the matter. In fact, the amplification may in some cases be so powerful as to produce a highly monochromatic flux of many megawatts/cm^2 with an input radiation of only several watts.

Two devices have been built around this phenomenon: the maser (microwave amplification by stimulated emission of radiation), and the laser (light amplification by stimulated emission of radiation). Although practically these devices are quite distinct, theoretically they are similar in that both are concerned with the amplification of electromagnetic radiation by stimulated emission of atomic radiation. The maser, as its name indicates, operates in the microwave region; the laser operates in the optical region (i.e. in the frequency range of 10^{14} to 10^{15} cycles per second).

2.13 Diatomic Molecules

Diatomic molecules with non-identical nuclei We consider in this subsection a perfect gas of diatomic molecules with nonidentical nuclei. Since the gas is perfect, the only interactions which arise are those internal to the molecules.

A diatomic molecule consists of two nuclei and a number of electrons which maintain the distance between the nuclei approximately constant. For low energies, the total internal energy of a molecule will be well approximated by the sum of the vibrational energy of the nuclei along their axis, and of the energy due to rotation of this axis. It turns out that the energy due to the moment of inertia of the electrons when the axis of the atom rotates is completely negligible at normal temperatures To within an additive constant, the internal energy of a diatomic molecule will be given by*

$$\varepsilon = n\hbar\omega + \frac{J(J+1)}{2I}\hbar^2, \qquad (J, n = 0, 1, 2, \ldots), \tag{2.84}$$

* See e.g. A. R. Edmonds, *Angular Momentum in Quantum Mechanics*, Princeton University Press (1960).

where I is the moment of inertia of the molecule, J its angular momentum, and ω its vibration frequency. There will, of course, be $(2J+1)$ states of the molecule having an angular momentum J.

The internal partition function

$$q_{\text{int}} = \sum e^{-\varepsilon/kT}$$

(the sum is over the energy levels enumerated in (2.84)), corresponding to the internal energy, can be written in the product form

$$q_{\text{int}} = q_{\text{rot}} \times q_{\text{vib}},$$

where

$$q_{\text{vib}} = \sum_{n=0}^{\infty} \exp(-n\hbar\omega/kT), \tag{2.85}$$

$$q_{\text{rot}} = \sum_{J} (2J+1) \exp[-J(J+1)\hbar^2/2IkT]. \tag{2.86}$$

The vibrational partition function q_{vib} can be easily evaluated. For normal temperatures, $(\hbar\omega/kT \gg 1)$ we have

$$q_{\text{vib}} = \frac{1}{1-e^{-\hbar\omega/kT}}. \tag{2.87}$$

For very high temperatures, it will be sufficient to consider only the first few terms in (2.85).

In order to evaluate the rotational partition function, we shall consider two limiting cases. We put

$$\gamma \equiv \hbar^2/2IkT,$$

$$y \equiv J(J+1)\gamma. \tag{2.88}$$

(i) when $\gamma \ll 1$ (large T) we have, taking the increment of (2.88)

$$\Delta y = (2J+1)\gamma\,\Delta J.$$

Going over from the sum in (2.86) to an integral, we have, since $\Delta J = 1$,

$$q_{\text{rot}} = \frac{1}{\gamma}\int_0^{\infty} e^{-y}\,dy = \frac{1}{\gamma}.$$

(ii) when $\gamma \gg 1$ (small T) we can retain only the first few terms in (2.86),

$$q_{\text{rot}} = [1+3e^{-2\gamma}+5e^{-6\gamma}+\ldots]. \tag{2.89}$$

Taking into account the translational motion of the molecules

(2.11) the partition function is

$$q = \frac{1}{N!}\left(\frac{V}{\lambda^3} q_{\text{int}}\right)^N.$$

The Helmholtz potential is then

$$A = -kT \ln q = -NkT \ln\left[\frac{V}{N}\left(\frac{2\pi\mu kT}{h^2}\right)^{\frac{3}{2}}\right] + A_{\text{rot}} + A_{\text{vib}},$$

where μ is the reduced mass of the molecule and

$$A_{\text{rot}} = -NkT \ln q_{\text{rot}},$$

$$A_{\text{vib}} = -NkT \ln q_{\text{vib}}.$$

Using the above relations, we obtain the following thermodynamic functions:

$$A_{\text{rot}} = \begin{cases} NkT \ln(2IkT/\hbar^2) & (\gamma \ll 1) \\ -NkT[\ln(1+3e^{-2\gamma}+\ldots)] & (\gamma \gg 1) \end{cases}$$

$$S_{\text{rot}} = \begin{cases} -Nk \ln(2IkT/\hbar^2) + Nk & (\gamma \gg 1) \\ \left(\dfrac{3N\hbar^2}{IJ} + 3Nk\right) e^{-\hbar^2/IkT} & (\gamma \gg 1) \end{cases}$$

$$S_{\text{vib}} = -Nk \ln(1 - e^{-\hbar\omega/kT}) + \frac{N\hbar\omega/T}{(e^{\hbar\omega/kT} - 1)}$$

$$E_{\text{vib}} = \frac{N\hbar\omega}{e^{\hbar\omega/kT} - 1} \cong \begin{cases} NkT & (T \gg \hbar\omega/k) \\ N\hbar\omega e^{-\hbar\omega/kT} & (T \ll \hbar\omega/k) \end{cases} \tag{2.90}$$

$$E_{\text{rot}} = \begin{cases} NkT & (\gamma \ll 1) \\ (3N\hbar^2/I)\, e^{-\hbar^2/2IkT} & (\gamma \gg 1) \end{cases} \tag{2.91}$$

For very high temperatures, the total energy is obtained from (2.89), (2.90) and (2.91):

$$E = \frac{3}{2}NkT + NkT + NkT = \frac{7}{2}NkT.$$

Hence

$$c_V = \frac{7}{2}Nk. \tag{2.92}$$

Since in a diatomic molecule there are three degrees of freedom corresponding to the translational modes of the centre of mass of the molecule, two degrees of freedom corresponding to the kinetic and potential energies of vibration, and two degrees of freedom corresponding to the rotation of the molecular axis, the result (2.92) agrees with theorem of equipartition of energy (section 2.5).

The constant $\hbar^2/2Ik$ varies from 85°K for the hydrogen molecule to 0·9°K for the iodine molecule. The constant $\hbar\omega/k$ varies from 5958°K for H_2 to 305°K for I_2. These are the extreme values that the constants assume. All other diatomic molecules have constants with values intermediate between these. From (2.85) and (2.86) therefore, it follows that as the temperature rises, first the rotational energy becomes important, then the vibrational energy takes on importance.

Diatomic molecules with identical nuclei: ortho and parahydrogen

When a gas consists of diatomic molecules with identical nuclei, symmetry effects must be considered. Let us take as an example the hydrogen molecule whose nuclei are identical. There exist two possible states of the molecule. If the spins of the nuclei are parallel, the total spin S of the molecule will be equal to I, and the molecule is called an *orthohydrogen* molecule. If, on the other hand, the spins of the nuclei are anti-parallel, then the total spin of the molecule will vanish ($S = 0$) and the molecule is called a *parahydrogen* molecule. Since the spin state of orthohydrogen is symmetrical, the spatial part of its wave function must be antisymmetrical. Similarly, since the spin state of parahydrogen is antisymmetrical, its spatial wave function must be symmetrical.

It follows that the angular momentum J of orthohydrogen (parahydrogen) can only take on odd (even) values. The corresponding rotational partition functions of the molecules will therefore be

$$q_{\rm rot}(\text{ortho}) = 3 \sum_{J=1,3,5,\ldots} (2J+1)\, e^{-J(J+1)\gamma}, \tag{2.93}$$

$$q_{\rm rot}(\text{para}) = 1 \sum_{J=0,2,4\ldots} (2J+1)\, e^{-J(J+1)\gamma} \tag{2.94}$$

The multiplicative factors 3 and 1 in (2.93) and (2.94) come from the spin degeneracies $(2S+1)$. The above expressions should be compared with the expression (2.86) for molecules with non-identical nuclei.

When $\gamma \ll 1$, one has, since now J varies by steps of 2 ($\Delta J = 2$):

$$q_{\rm rot}(\text{ortho}) = \frac{3}{2\gamma}\int_0^\infty e^{-y}dy = \frac{3}{2\gamma},$$

$$q_{\mathrm{rot}}\,(\mathrm{para}) = \frac{1}{2\gamma}.$$

When $\gamma \gg 1$,

$$q_{\mathrm{rot}}\,(\mathrm{ortho}) = 3(3e^{-2\gamma}+\ldots) = 9e^{-2\gamma}+\ldots, \tag{2.95}$$

$$q_{\mathrm{rot}}\,(\mathrm{para}) = [1+5\mathrm{e}^{-6\gamma}+\ldots]. \tag{2.96}$$

From (2.95) and (2.96) one can obtain the rotational energies of ortho- and parahydrogen at low temperatures

$$E_{\mathrm{rot}}\,(\mathrm{para}) \cong 15N(\mathrm{para})(\hbar^2/I)\,e^{-3\hbar^2}/IkT,$$

$$E_{\mathrm{rot}}\,(\mathrm{ortho}) \cong N(\mathrm{ortho})\,\hbar^2/\mathrm{I},$$

where N(para) and N (ortho) are the numbers of molecules of para- and orthohydrogen in the gas.

When the gas is in thermodynamic equilibrium, the partition function for the gas will be the sum of the partition functions for the ortho and parahydrogen:

$$q_{\mathrm{rot}} = q_{\mathrm{rot}}\,(\mathrm{ortho})+q_{\mathrm{rot}}\,(\mathrm{para}). \tag{2.17}$$

The relative number of molecules that are orthohydrogenic and parahydrogenic are given by the ratio of the Boltzmann distributions summed over the relevant states and multiplied by the appropriate statistical factors. This gives the ratio of the corresponding partition functions

$$\frac{N(\mathrm{ortho})}{N(\mathrm{para})} = \frac{3 \sum\limits_{J=1,3,5\ldots} (2J+1)\,e^{-J(J+1)\gamma}}{\sum\limits_{J=0,2,4\ldots} (2J+1)\,e^{-J(J+1)\gamma}}. \tag{2.98}$$

It is apparent from this ratio that since parahydrogen has a lower energy than orthohydrogen at very low temperatures, a gas of hydrogen molecules in thermodynamic equilibrium will be essentially a gas of parahydrogen. For higher temperatures.

$$\gamma \ll 1,$$

the ratio will tend towards the numerical value of 3.

The more common situation, however, is one of non-equilibrium, where the ratio N(ortho)/(N(para) does not vary continuously as in (2.98). This is because the quantum mechanical transition probabilities for the exchanges

$$\mathrm{ortho} \rightleftarrows \mathrm{para}$$

are very small. Therefore, as the temperature is lowered, the 3 to 1 ratio between ortho and parahydrogen will persist and all thermodynamic functions can be calculated from (2.97) where, however, N(ortho) and N(para) are given at the start.

The above considerations apply also to the case of the deuterium molecule D_2. For other molecules, the ortho-para distinction is either unimportant or non-existent.

EXERCISES

1 Consider a perfect classical gas in two dimensions. Find
(a) the partition function of the gas
(b) its entropy
(c) its internal energy
(d) the equation of state

2 Calculate the energy of a perfect classical gas using the microcanonical distribution.

3 Two identical particles of mass m interact each with an external harmonic oscillator potential, but do not interact with each other. Thus the Hamiltonian for the system is

$$H = \frac{1}{2m}p_1^2 + \frac{1}{2m}p_2^2 + \frac{1}{2}m\omega^2(x_1^2 + x_2^2)$$

where the frequency ω is a constant, and where the energy spectrum of each particle is given by the expression

$$E_n = \hbar\omega(n + \tfrac{1}{2}) \quad n = 0, 1, 2, \ldots$$

Find the specific heat at constant volume of the system when the particles are
(a) fermions
(b) bosons
(c) distinguishable particles

4 Find a proper generalization of the equipartition theorem applicable to relativistic particles.

5 Find the mean occupation number of a Boltzmann gas using the Darwin–Fowler method.

6 Could one observe an Einstein condensation in a photon gas? Explain.

7 Could one observe an Einstein condensation in a gas of bosons constrained to move on a one-dimensional or two-dimensional surface?

8 Suppose that in our treatment of Einstein condensation one counted as discrete states not only the ground state, but also the first excited state. Would one then obtain the same equation of state for temperatures below the critical temperature?

9 Calculate the fluctuations in particle number of an ideal boson gas for temperatures below the critical temperature for condensation. What peculiarities are displayed?

10 Show how the Planck radiation law is modified when electromagnetic radiation is in equilibrium in a cavity with a material having an index of refraction n.

11 Find the temperature T of a black body which emits a maximum amount of radiation at a wavelength $\lambda = 6929$ Å. Find the total radiation of this black body and the amount of radiation that it emits in the interval

$$(6929-0{\cdot}05)\,\text{Å} < \lambda < (6929+0{\cdot}05)\,\text{Å}$$

12 The potential energy of a classical anharmonic oscillator is given by the expression

$$V(x) = ax^2 - bx^3 - cx^4$$

where a, b, c are positive constants. Find the specific heat of the oscillator to order T, where T is the temperature.

CHAPTER 3

The Imperfect Classical Gas

3.1 Introduction

In this chapter we shall consider imperfect gases, i.e. gases for which the interactions between the constituent particles are no longer negligible.

We will set up a scheme that will alow us to calculate, by successive approximations, deviations from the perfect gas laws. Each approximation will determine a coefficient in the virial expansion of the pressure of the gas as a function of its density.

The problem of arriving at an equation of state which correctly describes the condensation phenomenon is a much more arduous one and is still a subject of discussion. The main question is how, starting from a continuous partition function, does one obtain a P–V curve which, at least for temperatures below a certain critical temperature, must display three different analytical parts, corresponding to the vapour, liquid, and mixed phases.

We recall that in the discussion of Einstein condensation, the thermodynamic limit played an important role, and we can well believe that this will also be true here.

We will not enter into a detailed discussion of the condensation phenomenon. However, we will introduce a historically important equation, the Van der Waals' equation, which can be obtained by interpolation from the low-density behaviour of real gases, and which, when combined with the so-called Maxwell equal-area rule, gives a qualitatively correct description of condensation.

The Hamiltonian for the gas will be

$$H = T + \sum_{i>j} u_{ij}, \tag{3.1}$$

where T is the total kinetic energy and u_{ij} the potential energy between the particles i and j. We shall suppose that u_{ij} depends only upon the distance between the pair of particles i and j,

$$r_{ij} \equiv |\vec{r}_i - \vec{r}_j|.$$

For a classical imperfect gas, the partition function can be written as a product

$$q = q_{\text{trans}} \cdot \frac{q_{\text{int}}}{V^N}, \tag{3.2}$$

where q_{trans} is the contribution to q arising from the kinetic energy of the gas, and where

$$q_{\text{int}} = \int \exp[-\sum_{i>j} u_{ij}/kT] \prod_{i=1}^{N} d^3\vec{r}_i \tag{3.3}$$

results from the interactions between all pairs of particles i and j induced by the potential u_{ij}. In (3.3) each coordinate can vary within the entire volume V.

We introduce the function

$$f_{ij} \equiv \exp(-u_{ij}/kT) - 1 \tag{3.4}$$

In the limit, as a gas approaches ideality, $f_{ij} \to 0$, and thus f_{ij} is a measure of the degree of 'imperfection' of the gas.

We have

$$\exp[-\sum_{i>j} u_{ij}/kT] = \prod_{i>j} (1 + f_{ij})$$

$$= 1 + \sum_{i>j} f_{ij} + \sum_{\substack{i>j \\ k>l \\ i,j \neq k,l}} f_{ij} f_{kl} + \ldots \tag{3.5}$$

Suppose now that one expands the exponential in (3.3) as in (3.5). Of course this expansion converges since the number of terms in (3.5) is finite. From a practical point of view, however, such an expansion would be useful only if the successive terms in the corresponding expansion of q_{int} (containing one term for each summation in (3.5)) became smaller and smaller, at least for low densities. To see whether this is so, let us calculate the order of magnitude of the contributions to (3.2) from the first two terms in (3.5).

The first term, when inserted in (3.3), gives the contribution of a perfect gas: V^N. To evaluate the second term we introduce the coordinates

$$\vec{r}_{ij} = \vec{r}_i - \vec{r}_j \qquad \vec{R} = \tfrac{1}{2}(\vec{r}_i + \vec{r}_j).$$

We find

$$\sum_{i>j} \int f_{ij}(r_{ij}) \prod_{k=1}^{N} d^3\vec{r}_k = V^{N-2} \sum_{i>j} \int f_{ij}(r_{ij})\, d^3\vec{r}_i\, d^3\vec{r}_j$$

$$= V^{N-1} \sum_{i>j} \int f(r_{ij})\, d^3\vec{r}_{ij}. \tag{3.6}$$

The integration over $\vec{R}$ gave an additional factor of V. The last integrals are taken over finite V-dependent regions. We shall assume here and in similar integrals that we can extend these finite regions of integration to cover all of space. The error introduced by doing this appears in the form of corrective terms which become smaller as V increases. We shall make the hypothesis that such corrective terms do not contribute to the partition function in the thermodynamic limit. Henceforth, we shall drop them.

Since within a group of N particles, there are $N(N-1)/2$ pairs of particles, the last term above contributes to the partition function a factor

$$\tfrac{1}{2}N(N-1)V^{N-1}\alpha \tag{3.7}$$

where

$$\alpha \equiv \int f(r_{ij})\, d^3\vec{r}_{ij}$$

is a constant. The ratio of (3.7) to V^N is of the order of

$$N(N-1)\alpha/2V \approx \tfrac{1}{2}N\alpha\rho,$$

where $\rho = N/V$. Hence the contribution to the partition function of the second term in (3.5) is of order N times larger than the contribution of the first term. The expansion obtained by substituting (3.5) into (3.2), as it stands, is therefore not useful for practical calculations.

However, since the series (3.5) converges, the terms can be rearranged to obtain a useful result. A method due to N. Van Kampen, which in effect amounts to a rearranging of terms, will be described in the following section.

3.2 The Method of Van Kampen

We introduce the function

$$\varphi_{ij} \equiv \exp(-u_{ij}/kT),$$

which is related to f_{ij} (*cf.* 3.4) by the relation

$$\varphi_{ij} = f_{ij}+1. \tag{3.8}$$

Since each particle can move within the entire volume V, the average taken over all configurations of the product $\varphi_{12}\varphi_{13}\cdots\varphi_{N-1,N}$ is

$$\overline{\varphi_{12}\varphi_{13}\cdots\varphi_{N-1,N}} = \frac{\int \exp[-\sum_{i>j} u_{ij}/kT] \prod_{i=1}^{N} d^3\vec{r}_i}{\int \prod_{i=1}^{N} d^3\vec{r}_i}$$

$$= \frac{\int \exp[-\sum_{i>j} u_{ij}/kT] \prod_{i=1}^{N} d^3\vec{r}_i}{V^N} \tag{3.9}$$

$$= \frac{q_{\text{int}}}{V^N}$$

Let us note an important property of φ_{ij}. When the distance r_{ij} between the particles i and j is greater than the range of the interparticle potential, one has

$$\varphi_{ij} \to 1.$$

Now any two φ_{ij} (for example φ_{12} and φ_{13}) are always statistically independent in the sense that their product is factorizable

$$\overline{\varphi_{12}\varphi_{13}} = \overline{\varphi}_{12}\overline{\varphi}_{13} = (\overline{\varphi}_{12})^2 = (\overline{\varphi}_{13})^2. \tag{3.10}$$

This can be easily proved by introducing the coordinates

$$\vec{R} = \tfrac{1}{3}(\vec{r}_1+\vec{r}_2+\vec{r}_3), \quad \vec{r}_{12} = \vec{r}_2-\vec{r}_1, \quad \vec{r}_{13} = \vec{r}_3-\vec{r}_1$$

We then have

$$\overline{\varphi_{12}\varphi_{13}} = \frac{1}{V^3}\int \varphi_{12}\varphi_{13}\, d^3R d^3r_{12}\, d^3r_{13}$$

$$= \frac{1}{V^2}\left[\int \varphi_{12}\, d^3r_{12}\right]\left[\int \varphi_{13}\, d^3r_{13}\right]$$

$$= \overline{\varphi_{12}}\,\overline{\varphi_{13}} = (\overline{\varphi_{12}})^2 = (\overline{\varphi_{13}})^2.$$

The relation (3.10) can be interpreted as follows. The probabilities that the pairs of particles 1 and 2, and also 1 and 3 are both within the range of the potential, are independent.

The statistical independence of a product of three φ_{ij} is not, however, ensured. That is to say, in general

$$\overline{\varphi_{12}\varphi_{13}\varphi_{23}} \neq \overline{\varphi_{12}}\,\overline{\varphi_{13}}\,\overline{\varphi_{23}}. \tag{3.11}$$

(i) *The First Approximation* When the density of the gas is very low, it is not very probable that among a group of three particles, all three particles will be simultaneously within the range of the interparticle potential. On the contrary, because the gas is rarified, it is much more probable that at least one of the functions φ_{12}, φ_{13}, φ_{23} will be equal to unity. In this case, a product of three functions φ_{ij} is practically reduced to a product of two functions, which, as we have just seen, is statistically independent. In a first approximation we shall therefore write

$$\overline{\varphi_{12}\varphi_{13}\varphi_{23}} \cong \overline{\varphi_{12}}\,\overline{\varphi_{13}}\,\overline{\varphi_{23}} = (\overline{\varphi_{12}})^3.$$

The above approximation means that in each group of three particles, only two-particle correlations are taken into account. If in a group of N particles, we take into account only two-particle correlations, we will have

$$\overline{\varphi_{12}\varphi_{13}\cdots\varphi_{N-1,N}} \cong \overline{\varphi_{12}}\,\overline{\varphi_{13}}\cdots\overline{\varphi_{N-1,N}}$$

$$= (\overline{\varphi_{12}})^{\frac{1}{2}N(N-1)}$$

$$= \left[\left(\frac{V^{N-2}}{V^N}\right)\int \exp(-u(r_{12})/kT)\,d^3r_1\,d^3r_2\right]^{\frac{1}{2}N(N-1)}$$

$$= \left[\frac{1}{V}\int \exp(-u(r_{12})/kT)\,d^3r_{12}\right]^{\frac{1}{2}N(N-1)}$$

$$\equiv \left[1+\frac{1}{V}\int (\exp(-u(r_{12})/kT)-1)\,d^3r_{12}\right]^{\frac{1}{2}N(N-1)}.$$

Comparing with (3.9) one finds in the thermodynamic limit for the first approximation $q_{\text{int}}^{(1)}$ to q_{int}

$$\lim_{\substack{N\to\infty \\ V\to\infty \\ \rho=\text{const.}}} \left[\frac{q_{\text{int}}^{(1)}}{V^N}\right]^{1/N} = \lim_{\substack{N\to\infty \\ V\to\infty \\ \rho=\text{const.}}} \left[1+\frac{1}{V}\int \{\exp(-u(r_{12})/kT)-1\}\, d^3r_{12}\right]^{\frac{1}{2}(N-1)} \tag{3.12}$$

$$= \exp\left[\frac{N}{2V}\int \{\exp(-u(r_{12})/kT)-1\}\, d^3r_{12}\right].$$

Put

$$\beta_1 \equiv \int \{\exp(-u(r_{12})/kT)-1\}\, d^3r_{12} \equiv \int f(r_{12})\, d^3r_{12}. \tag{3.13}$$

β_1 is called the *irreducible cluster integral of order* 1.* From (3.12) one obtains the formula

$$q_{\text{int}}^{(1)} = V^N \exp(\tfrac{1}{2}\rho N\beta_1 + 0(1)), \tag{3.14}$$

which tends to V^N when $\beta_1 \to 0$, (the perfect gas limit).

(ii) *The Second Approximation* In second approximation it is natural to include up to three-particle correlations. For each group of three particles, the approximation to the partition function q_{int} will then have to be corrected by a multiplicative factor of the type

$$\overline{\varphi_{12}\varphi_{13}\varphi_{23}}/(\overline{\varphi_{12}})^3.$$

Using (3.8), this can be written

$$\frac{\overline{\varphi_{12}\varphi_{13}\varphi_{23}}}{(\overline{\varphi_{12}})^3} = \frac{[1+3\overline{f_{12}}+3(\overline{f_{12}})^2+\overline{f_{12}f_{13}f_{23}}]}{[1+3\overline{f_{12}}+3(\overline{f_{12}})^2+(\overline{f_{12}})^3]}. \tag{3.15}$$

But

$$\overline{f_{12}} = \frac{V^{N-2}}{V^N}\int \{\exp(-u(r_{12})/kT)-1\}\, d^3r_1\, d^3r_2$$

$$= \frac{1}{V}\int \{\exp(-u(r_{12})/kT)-1\}\, d^3r_{12} = 0\left(\frac{1}{V}\right).$$

* The irreducible cluster integrals should not be confused with the reducible cluster integrals introduced by Ursell in his method of treating the imperfect gas. These integrals are in general not equal. For a clear account of Ursell's method see e.g. K. Huang, *Statistical Mechanics*, John Wiley and Sons, Inc., New York (1965).

Also

$$\overline{f_{12}f_{13}f_{23}} = \frac{V^{N-3}}{V^N}\int[\{\exp(-u(r_{12})/kT)-1\}\{\exp(-u(r_{13})/kT)-1\}$$
$$\times\{\exp(-u(r_{23})/kT)-1\}\,d^3r_1\,d^3r_2\,d^3r_3\,]$$
$$= \frac{1}{V^2}\int[\{\exp(-u(r_{12})/kT)-1\}\{\exp(-u(r_{13})/kT)-1\}$$
$$\times\{\exp(-u(|\vec{r}_{13}-\vec{r}_{12}|)/kT)-1\}\,d^3r_{12}\,d^3r_{13}\,]$$
$$= 0(1/V^2).$$

Thus by expanding (3.15) in powers of $1/V$ one finds, for very large V

$$\frac{\overline{\varphi_{12}\varphi_{13}\varphi_{23}}}{(\overline{\varphi_{12}})^3} = \left[1+\overline{f_{12}f_{13}f_{23}}+0\left(\frac{1}{V^3}\right)\right]. \tag{3.16}$$

But there are

$$\binom{N}{3} = \frac{N(N-1)(N-2)}{6}$$

triplets among N particles. Therefore the three-particle correlations correct q_{int} by the factor

$$[1+\overline{f_{12}f_{13}f_{23}}+0(1/V^3)]^{N(N-1)(N-2)/6}.$$

We define the irreducible cluster integral of order 2 as

$$\beta_2 = \tfrac{1}{2}\int[\{\exp(-u(\vec{r})/kT)-1\}\{\exp(-u(\vec{r}\,')/kT)-1\}$$
$$\times\{\exp(-u(\vec{r}-\vec{r}\,')/kT)-1\}\,d^3\vec{r}\,d^3\vec{r}\,'].\tag{3.17}$$

so that

$$\beta_2/V^2 = \tfrac{1}{2}\overline{f_{12}f_{13}f_{23}}+0(1/V^3).$$

The contribution from three-particle correlations corrects $q_{\text{int}}^{(1)}$ by the factor

$$q_{\text{int}}^{(2)} = \left[1+\frac{2}{V^2}\beta_2+0\left(\frac{1}{V^3}\right)\right]^{N(N-1)(N-2)/6}.$$

In the thermodynamic limit one has

$$\lim_{\substack{N\to\infty\\ V\to\infty}} [q_{\text{in}}^{(2)}]^{1/N} = \exp\{\tfrac{1}{3}\rho^2\beta_2\},$$

whence

$$q_{\text{int}}^{(2)} = \exp\{\tfrac{1}{3}N\rho^2\beta_2 + 0(1)\}. \tag{3.18}$$

In second approximation (3.14) and (3.18) give the result

$$q_{\text{int}} = V^N \exp\{\tfrac{1}{2}\rho N\beta_1 + \tfrac{1}{3}\rho^2 N\beta_2 + 0(1)\}. \tag{3.19}$$

(iii) *The General Case* We wish now to generalize the preceding results and find the term of the nth order in the density in the expansion of the partition function. Thus we shall consider correlations involving up to $n+1$ particles and calculate the ratio

$$\overline{\varphi_{12}\varphi_{13}\cdots\varphi_{n,n+1}}/\Delta, \tag{3.20}$$

where Δ is the contribution of the average $\overline{\varphi_{12}\varphi_{13}\cdots\varphi_{n,n+1}}$ to the term of order $n-1$ in the density. The denominator Δ will thus contain up to n particle correlations.

As before we express the φ_{ij} in terms of the f_{ij} and obtain for (3.20)

$$\frac{1}{\Delta}\left\{1 + \sum_{i<j}\overline{f_{ij}} + \sum_{\substack{i<j\\ k<l}}\overline{f_{ij}f_{kl}} + \ldots + \overline{f_{12}f_{13}f_{14}\cdots f_{n-1,n}}\right\}. \tag{3.21}$$

Before proceeding further it is convenient to introduce a definition. We shall say that a given term in (3.21) (which, typically, is the average of a product of f_{ij} involving, say, k particles) is *reducible* if it can be factored into distinct products of averages, each factor involving fewer that k particles. Obviously such terms will describe correlations among fewer than k particles. If a term is not factorizable, we shall say that it is *irreducible*.

Example The averages of the products $f_{12}\,f_{13}$ and $f_{12}\,f_{13}\,f_{23}\,f_{14}$ are reducible since

$$\overline{f_{12}\,f_{13}} = \overline{f_{12}}\,\overline{f_{13}},$$

$$\overline{f_{12}\,f_{13}\,f_{23}\,f_{14}} = \overline{f_{12}\,f_{13}\;{}_{23}}\cdot\overline{f_{14}}. \tag{3.22}$$

On the other hand the term $\overline{f_{12}\,f_{13}\,f_{23}}$ is irreducible (*cf.* 3.16). The proof of (3.22), which is left to the reader, can be carried out by transforming integration variables as in (3.16).

The following example will illustrate in some detail the notions just introduced

Example Consider the average

$$\overline{\varphi_{12}\varphi_{13}\varphi_{14}\varphi_{23}\varphi_{24}\varphi_{34}}$$

involving up to four-particle correlations. In terms of the f_{ij}, this can be written (*cf.* 3.21)

$$\begin{aligned}
&\overline{\varphi_{12}\varphi_{13}\varphi_{14}\varphi_{23}\varphi_{24}\varphi_{34}} \\
&= 1+\sum_{i<j}\overline{f_{ij}}+\sum_{\substack{i<j\\k<l}}\overline{f_{ij}f_{kl}} \qquad (3.23)\\
&+\sum_{i<j\,\text{etc.}}\overline{f_{ij}f_{kl}f_{mn}}+\sum_{i<j\,\text{etc.}}\overline{f_{ij}f_{kl}f_{mn}f_{rs}} \\
&+\sum_{i<j\,\text{etc.}}\overline{f_{ij}f_{kl}f_{mn}f_{rs}f_{uv}}+\overline{f_{12}f_{13}f_{14}f_{34}f_{23}f_{24}}.
\end{aligned}$$

The first sum on the RHS contains six similar terms. In the second sum, all $\binom{6}{2}=15$ products which involve two f's are obviously reducible (*cf.* 3.10). One can easily verify that among the $\binom{6}{3}=20$ products involving three f's in the third sum, 16 are reducible. Similarly, 12 of the $\binom{6}{4}=15$ products involving four f's are reducible, while all $\binom{6}{5}=6$ terms involving five f's are irreducible. Thus (3.23) becomes

$$\begin{aligned}
&\overline{\varphi_{12}\varphi_{13}\varphi_{14}\varphi_{23}\varphi_{24}\varphi_{34}} \\
&=1+6\overline{f_{12}}+15\overline{f_{12}}^2 \\
&+16\overline{f_{12}}^3+4\overline{f_{12}f_{13}f_{23}}+12\overline{f_{12}}\cdot\overline{f_{12}f_{13}f_{23}} \\
&+3\overline{f_{12}f_{14}f_{23}f_{34}}+6\overline{f_{12}f_{13}f_{23}f_{34}f_{14}} \\
&+\overline{f_{12}f_{13}f_{14}f_{23}f_{24}f_{34}}. \qquad (3.24)
\end{aligned}$$

Let us now return to our main task which is the evaluation of (3.20). Among all the terms appearing in (3.21), some will involve fewer than $n+1$ particles (for an illustration, see (3.24)). These terms, obviously, will be present in the denominator Δ. Among the terms involving exactly $n+1$ particles, those that are reducible will also appear in Δ since these terms describe correlations among fewer than $n+1$ particles.

It remains to consider the irreducible $(n+1)$-particle terms which, when they appear in the numerator, are evidently of order $1/V^n$. These same terms also appear in the denominator. There, however, they describe lower order correlations and are therefore at least of order $1/V^{n+1}$.

Let R and I denote, respectively, the reducible and irreducible terms in the numerator of (3.21). We have

$$\frac{\overline{\varphi_{12}\,\varphi_{13}\cdots\varphi_{n,n+1}}}{\Delta} = \frac{1+R+I}{1+R+0(V^{-n-1})}$$
$$= 1+I+0(V^{-n-1}).$$

Defining the *irreducible cluster integral of order n*, β_n, by the relation

$$n!\beta_n/V^N \equiv I = \sum \overline{f_{ij}\,f_{kl}\,f_{mn}\cdots}, \tag{3.25}$$

where the sum is taken over all irreducible terms involving $n+1$ particles, we find

$$\frac{\overline{\varphi_{12}\,\varphi_{13}\cdots\varphi_{n,n+1}}}{\Delta} = 1+\frac{n!}{V^n}\beta_n+0(V^{-n-1}).$$

Since, among N particles, there are, for large N

$$\binom{N}{n+1} \cong \frac{N^{n+1}}{(n+1)!}$$

groups of $(n+1)$ particles, one finds in the thermodynamic limit

$$\lim_{\substack{N\to\infty\\ V\to\infty}} [q_{\text{int}}^{(n)}]^{1/N} = \lim_{\substack{N\to\infty\\ V\to\infty}} [1+n!\beta_n/V^n+0(V^{-n-1})]^{N^n/(n+1)!}$$
$$= \exp\{\rho^n\,\beta_n/n+1\}.$$

Thus, finally, the interaction part of the partition function is

$$q_{\text{int}} = V^N \exp\left\{N\sum_{n=1}^{\infty}\frac{\rho^n}{n+1}\beta_n+0(1)\right\}, \tag{3.26}$$

and the equation of state is

$$\frac{P}{kT} = \frac{\partial}{\partial V}\ln\left(q_{\text{trans}}\frac{q_{\text{int}}}{V}\right)$$

$$= \rho - \sum_{n=1}^{\infty} \frac{n}{n+1}\beta_n \rho^{n+1}. \tag{3.27}$$

The coefficient of ρ^n in (3.27) is called the nth *virial coefficient.*

3.3 Van der Waals' Equation

As an application of the previous theory, we shall consider a gas consisting of hard spherical molecules of radius r_0

$$u(r) = +\infty \text{ for } r < r_0$$

We shall suppose that for $r > r_0$ there is a weak, long-range, intermolecular attraction. This, in fact, corresponds to the experimental situation.

The function f defined by (3.4) will be

$$f = \begin{cases} -1 & r < r_0 \\ e^{-u/kT} \approx -u/kT & r > r_0 \end{cases} \tag{3.28}$$

The irreducible cluster integral of order 1 is (cf. 3.13)

$$\beta_1 = \int f(r)\, d^3r = -\int_{r<r_0} d^3r - \frac{1}{kT}\int_{r>r_0} u(r)\, d^3\vec{r}. \tag{3.29}$$

We put

$$a \equiv -\int_{r>r_0} u(r)\, d^3\vec{r}\,; \quad b \equiv \int_{r<r_0} d^3\vec{r}. \tag{3.30}$$

Hence

$$\beta_1 = -b + \frac{a}{kT}.$$

We can now write (3.26) including terms up to the second virial coefficients, as

$$q_{\text{int}} = \left[V\exp\left(-\frac{bN}{V}\right)\right]^N \exp\left(\frac{\rho a}{kT}\right). \tag{3.31}$$

Equation (3.31) is an approximate formula inasmuch as only the first two virial coefficients have been considered. This means, practically, that the volume of the gas is large compared to the proper volume bN of all the molecules and that correlations of order higher than 2 can be neglected. However, it must be recognized that (3.31) is deficient in at least one respect: namely for a gas composed of hard spheres, q_{int} should go to zero as $V \to bN$ (see equation 3.9).

When $V \gg bN$, we can expand the exponential in the first factor in (3.31) and retain only the first two terms

$$V \exp(-bN/V) \to (V-bN).$$

If we now take the RHS of this equation as an interpolation formula for the LHS, we will also satisfy the above-mentioned requirement that $q_{\text{int}} \to 0$ as $V \to bN$.

Thus we replace (3.31) by

$$q_{\text{int}} \to (V-bN)^N \exp\left(\frac{N\rho a}{kT}\right),$$

and obtain for the equation of state

$$\frac{P}{kT} = \frac{\partial}{\partial V} \ln\left(\frac{q_{\text{trans}}}{V^N} q_{\text{int}}\right) = \frac{N}{V-bN} - \frac{\rho^2 a}{kT}$$

or

$$(P+\rho^2 a)(V-bN) = NkT, \tag{3.32}$$

which is Van der Waal's equation.

From the non-unique manner in which Van der Waals' equation was derived, we can hardly expect to have obtained an exact result. Van der Waals' equation in fact gives only a qualitatively correct description of the departure of gases from the perfect gas laws, and of condensation phenomena.

For large V the isotherms of Van der Waals' equation are those of a perfect gas. As the pressure and volume are lowered, there is a region where the P–V curve becomes triple-valued with respect to P (Fig. 2). This phenomenon signifies the appearance of a gas-liquid phase transition.

The substance will occupy the volume corresponding to the most stable situation, which means that in the liquid phase it will be at higher pressures and in the vapour phase at lower pressures. At some pressure P_0, both phases will be stable, and the isotherm on the P–V diagram will have a horizontal portion. To determine this pressure, consider the closed cycle $ABCDECA$ (Fig. 2). From the second law of thermodynamics, the total work done as the substance goes through this isothermal cycle must vanish

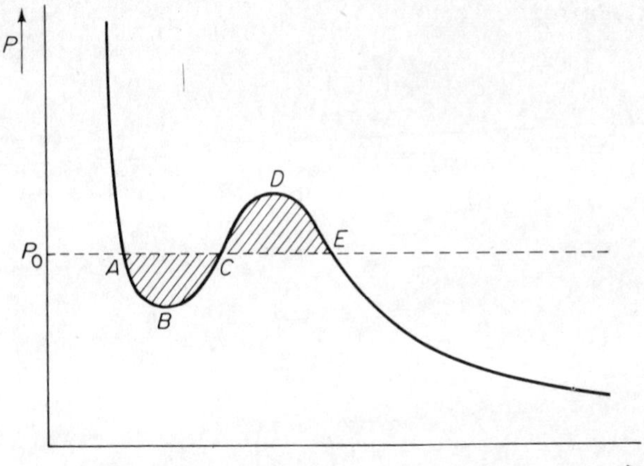

Fig. 2 An isotherm of a Van der Waals' gas

$$\oint V\,dP = 0. \tag{3.33}$$

Thus the two shaded areas of Fig. 2 must be equal. This result, known as *Maxwell's equal area rule*, determines P_0.

An objection which can be raised against Van der Waals' equation is that for small enough volumes, the compressibility of the gas

$$\kappa = -\frac{1}{V}\left(\frac{\partial V}{\partial P}\right)_T = \frac{1}{V}\left[\frac{NkT}{(V-bN)^2} - \frac{2N^2a}{V^3}\right]^{-1}$$

becomes negative. This, of course cannot happen for a real gas, and it has been shown rigorously that even for the model considered here this is not possible.* Still another objection to the equation is that the value of the coefficient b, calculated from (3.25), and which is equal to the effective volume of a gas molecule, does not give a value close to the experimentally measured one. In Van der Waals' equation coefficients a and b must be adjusted to fit the experiments.

However, it is possible to eliminate from Van der Waals' equation the coefficient a and b, and bring it to a more generally valid form.

In order to do this, we note that there exists a critical point where (3.32) has a triple root with respect to V. Let the coordinates of this critical point be denoted by V_c, P_c, and T_c. The critical point is then obtained from the condition that the inflexion point of the P–V

* D. Ruelle, *Helv. Phys. Acta,* **36**, (1963) 183.

curve is horizontal, i.e.

$$\left(\frac{\partial P}{\partial V}\right)_T = \left(\frac{\partial^2 P}{\partial V^2}\right)_T = 0,$$

which gives

$$V_c = 3bN; \quad T_c = \frac{8a}{27kb}; \quad P_c = \frac{a}{27b^2}.$$

Now putting

$$T' = \frac{T}{T_c}; \quad P' = \frac{P}{P_c}; \quad V' = \frac{V}{V_c}.$$

(3.32) becomes

$$\left(P' + \frac{3}{V'^2}\right)\left(V' - \frac{1}{3}\right) = \frac{8}{3}T'. \tag{3.34}$$

In (3.34) there appear no constants that characterize the molecule. This is a particular example of a *law of corresponding states* that is, a law expressing the fact that the equation of state for all members of some class of substances can be written in the form $P' = f(V', T')$ where the function f is the same for all substances in the group.

3.4 Calculation of the Third Virial Coefficient of a Hard Sphere Gas

We illustrate in this section the calculation of higher order virial coefficients. We shall find the irreducible cluster integral of order 2 (i.e. β_2), for a gas of hard spheres of diameter d:

$$\beta_2 = \tfrac{1}{2}V^2\,\overline{f_{12}\,f_{13}\,f_{23}} = \tfrac{1}{2}\int f_{12}\,f_{13}\,f_{23}\,d^3\vec{r}_{12}\,d^3\vec{r}_{13}. \tag{3.35}$$

One has

$$f_{ij} = \begin{cases} -0 & \text{if } r_{ij} < d, \\ 0 & \text{if } r_{ij} > d, \end{cases}$$

and the integral in (3.35) differs from zero only if the distances r_{12}, r_{13}, and r_{23} between the centres of the spheres obey the conditions

$$r_{12} < d \text{ and } r_{13} < d \text{ and } r_{23} < d. \tag{3.36}$$

If these three conditions are satisfied, then

$$f_{12}\,f_{13}\,f_{23} = -1.$$

so that

$$\beta_2 = -2\times\tfrac{1}{2}\int\limits_{\Omega} d^3\vec{r}_{12}\, d^3\vec{r}_{13},$$

where the integration region Ω is determined from the conditions (3.36). For a fixed distance $x(x < d)$ between the centres of the spheres 1 and 2, the centre of the third sphere can be located within a spatial region whose projection on the plane containing the centres of the

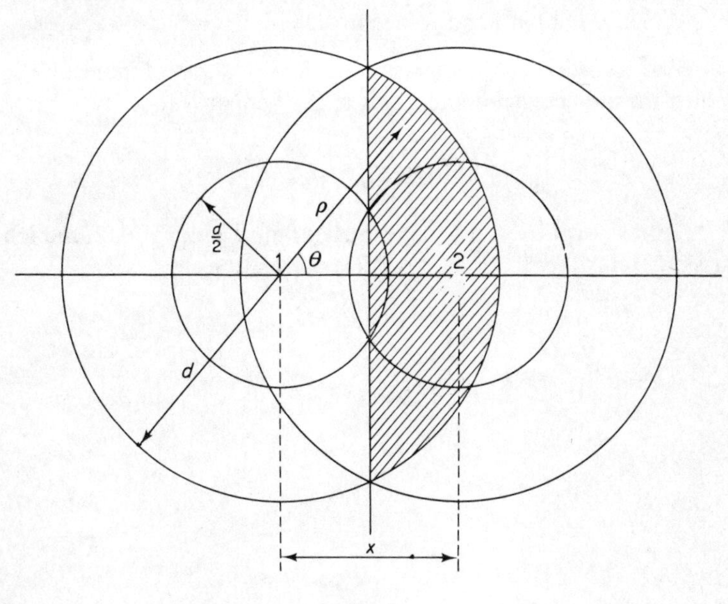

Fig. 3

fixed spheres is the shaded area of Fig. 3. Using the notations of Fig. 3, we have

$$\beta_2 = -(I_1 - I_2),$$

where

$$I_1 = \int_0^d 4\pi x^2\, dx \int_0^{\cos^{-1}(x/2d)} d\theta \int_0^{2\pi} d\varphi \int_0^d d\rho\, \rho^2 \sin\theta = \tfrac{5}{9}\pi^2 d^6,$$

$$I_2 = \int_0^d 4\pi x^2\, dx \int_0^{\cos^{-1}(x/2d)} d\theta \int_0^{2\pi} d\varphi \int_0^{x/(2\cos\theta)} d\rho\, \rho^2 \sin\theta$$

$$= 5\pi^2 d^6/36.$$

Thus

$$\beta_2 = -\tfrac{5}{12}\pi^2 d^6,$$

and the third virial coefficient of the hard sphere gas is

$$-\tfrac{2}{3}\beta_2 = +\tfrac{5}{18}\pi^2 d^6.$$

EXERCISES

1 Calculate the virial coefficient of order 3 of a gas composed of one-dimensional hard rods of length L.

2 Find to order ρ^2 the equation of state of a gas of molecules in which the molecular interaction is of the Lennard–Jones type

$$U(r) = \frac{A}{r^{12}} - \frac{B}{r^6}$$

3 Deduce explicitly the contribution to the partition function arising exclusively from four-particle correlations.

CHAPTER 4

Solids

4.1 Specific Heat of Solids: Introduction

The specific heat of a substance is the most important measurable thermodynamic quantity. Therefore we shall study it from the theoretical point of view in some detail.

At low temperatures, the specific heat of a solid consisting of N atoms, is smaller than its classical value $3Nk$ given by the theorem of equipartition of energy. For moderately low temperatures, Einstein has given an explanation of this phenomenon which, however, is no longer valid at extremely low temperatures. A more exact treatment which we shall discuss here is the one given by Born, Von Kármán and Debye.

The problem will be considered as follows. We shall quantize the elastic vibrations in a solid. These vibrations give rise to a 'sound' field whose quanta, the *phonons* as they are called, obey Bose–Einstein statistics. We shall deduce the dispersion law $E = f(k)$ which expresses the energy of a phonon in terms of its wave number, and this will allow us to calculate the energy density of the field and thus the specific heat of the solid.

Before attempting to solve this problem we must become familiar with a few notions of quantum field theory.

4.2 Second Quantization of the Harmonic Oscillator

Consider a harmonic oscillator of mass m. The Hamiltonian is given by the familiar expression

$$H = \frac{p^2}{2m} + \frac{m\omega^2}{2} q^2, \tag{4.1}$$

where q is the displacement from the equilibrium position, p is the momentum, and ω is the angular frequency. If p and q are operators which obey the commutation relation

$$[p, q] = \hbar/i, \tag{4.2}$$

then (4.1) will be the Hamiltonian of a quantized oscillator (this is called first quantization). We wish to diagonalize H. Put

$$q \equiv (a+a^+)\sqrt{\frac{\hbar}{2m\omega}}; \qquad p \equiv \frac{a-a^+}{i}\sqrt{\frac{\hbar m\omega}{2}}. \tag{4.3}$$

These relations define two operators a and a^+, and from (4.2) their commutation relation is given by

$$[a, a^+] = 1. \tag{4.4}$$

Using (4.3) and (4.4), one can express H in terms of a and a^+:

$$H = \tfrac{1}{2}\hbar\omega[aa^+ + a^+a] = \hbar\omega(a^+a+\tfrac{1}{2}). \tag{4.5}$$

But we know from quantum mechanics that the eigenvalues of a harmonic oscillator are

$$E_n = \hbar\omega(n+\tfrac{1}{2}), \qquad (n = 0, 1, 2, \ldots).$$

It follows that one can define an *occupation operator*

$$N = a^+a, \tag{4.6}$$

whose eigenvalues are the non-negative integers. The operators a and a^+ can be represented by matrices. For example, one could have

$$a = \begin{pmatrix} 0 & \sqrt{1} & 0 & 0 & 0 & . & . \\ 0 & 0 & \sqrt{2} & 0 & 0 & . & . \\ 0 & 0 & 0 & \sqrt{3} & . & . & . \\ . & . & . & . & . & . & . \\ . & . & . & . & . & . & . \end{pmatrix}$$

$$a^+ = \begin{pmatrix} 0 & 0 & 0 & 0 & 0 & . & . \\ \sqrt{1} & 0 & 0 & 0 & 0 & . & . \\ 0 & \sqrt{2} & 0 & 0 & 0 & . & . \\ 0 & 0 & \sqrt{3} & 0 & 0 & . & . \\ . & . & . & . & . & . & . \end{pmatrix}$$

and one would then obtain the diagonal matrix

$$N = a^+a = \begin{pmatrix} 0 & 0 & 0 & 0 & 0 & . & . \\ 0 & 1 & 0 & 0 & 0 & . & . \\ 0 & 0 & 2 & 0 & 0 & . & . \\ 0 & 0 & 0 & 3 & 0 & . & . \\ . & . & . & . & . & . & . \end{pmatrix}$$

representing the occupation operator.

Let $|n\rangle$ be a vector in Hilbert space representing a state which contains n quanta of the harmonic oscillator. By definition of $|n\rangle$, we will have

$$N|n\rangle = n|n\rangle. \tag{4.7}$$

We shall show that a^+ and a are, respectively, *creation* and *annihilation* operators for the quantas in the sense that

$$a^+|n\rangle \sim |n+1\rangle,$$

$$a|n\rangle \sim |n-1\rangle.$$

That is, a^+ increases and a decreases the occupation number of a state by one unit. One has the following commutation relations, which follow immediately from (4.4), (4.6) and (4.7):

$$[a^+, N] = -a^+, \tag{4.8}$$

$$[a, N] = +a. \tag{4.9}$$

Applying (4.8) to a state $|n\rangle$,

$$(a^+N - Na^+)|n\rangle = na^+|n\rangle - Na^+|n\rangle = -a^+|n\rangle.$$

whence

$$N[a^+|n\rangle] = (n+1)[a^+|n\rangle]. \tag{4.10}$$

Similarly, by applying (4.9) to a state $|n\rangle$, we obtain

$$(aN - Na)|n\rangle = na|n\rangle - Na|n\rangle = a|n\rangle.$$

Hence

$$N[a|n\rangle] = (n-1)[a|n\rangle]. \tag{4.11}$$

The relations (4.10) and (4.11) show that $a^+|n\rangle$ contains $n+1$ quanta whereas $a|n\rangle$ contains only $n-1$ quanta. This proves that

a^+ and a are, respectively, creation and annihilation operators for the quanta of the oscillator. The introduction of such operators is called *second quantization*.

4.3 One-Dimensional Model of a Solid: Acoustic Waves

In this section we will develop the theory of vibrations in a solid. This will lead us to the notion of phonons, which have some analogies with the photons of electromagnetic theory.

For our one-dimensional model of a solid, we will assume that we have a continuous elastic string which can vibrate along its longitudinal direction, and which can be represented by N equidistant points of mass M, connected together by massless springs (Fig. 4). The Nth particle will be assumed to interact with the first

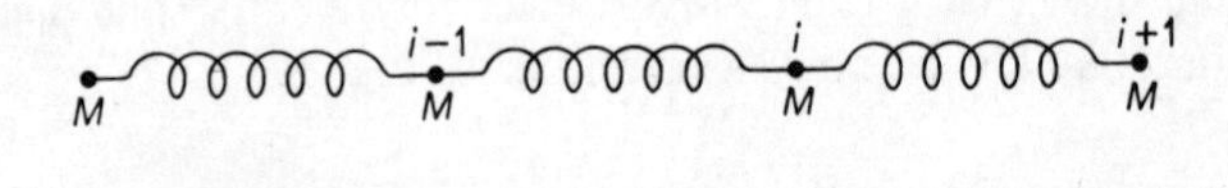

Fig. 4 A one-dimensional model of a solid

particle, and if q_i is the displacement of the ith mass from its equilibrium position, we will take

$$q_{N+1} \equiv q_1,$$

whence

$$q_{i+N} = q_i.$$

Our first concern will be to obtain the normal coordinates of the system by a purely classical treatment. The quantization of our model will come in at a later stage.

If $\mathscr{K}$ denotes the spring constant, then the force exerted on the ith mass will be

$$F_i = \mathscr{K}(q_{i+1}-q_i)-\mathscr{K}(q_i-q_{i-1}).$$

The first term represents the force exerted towards the right by the ith spring, and the second term represents the force exerted towards the left by the $(i-1)$th spring.

The potential energy is

$$V = \tfrac{1}{2}\mathscr{K}\sum_i (q_{i+1}-q_i)^2, \tag{4.12}$$

and the Lagrangian of the system is therefore

$$L = \tfrac{1}{2}M \sum_{i=1}^{N} \dot{q}_i^2 - \tfrac{1}{2}\mathscr{K} \sum_{i=1}^{N} (q_{i+1} - q_i)^2. \tag{4.13}$$

Letting p_i be the momentum of the ith particle, we have

$$p_i = \partial L/\partial \dot{q}_i = M\dot{q}_i.$$

The Hamiltonian is

$$H = \sum_i p_i \dot{q}_i - L = \frac{1}{2M} \sum_{i=1}^{N} p_i^2 + \frac{\mathscr{K}}{2} \sum_{i=1}^{N} (q_{i+1} - q_i)^2. \tag{4.14}$$

The potential energy may be written as

$$V = \sum_{i,j=1}^{N} q_i \mathscr{V}^i{}_j q_j, \tag{4.15}$$

where $\mathscr{V}_i^j$ $(i, j = 1, 2, \ldots, N)$ are the elements of a real, symmetric matrix

$$\mathscr{V} = \frac{\mathscr{K}}{2} \begin{pmatrix} 2 & -1 & 0 & 0 & 0 \ldots 0 & 0 & -1 \\ -1 & 2 & -1 & 0 & 0 \ldots 0 & 0 & 0 \\ 0 & -1 & 2 & -1 & 0 \ldots 0 & 0 & 0 \\ \cdot & & & & & \cdot & \\ \cdot & & & & & \cdot & \\ \cdot & & & & & \cdot & \\ -1 & 0 & 0 & 0 & 0 \ldots 0 & -1 & 2 \end{pmatrix} \tag{4.16}$$

Notice that in order to diagonalize the Hamiltonian, it is sufficient to diagonalize the potential energy.

The method required to express the quadratic form (4.15) as a sum of squares is well known. One needs to find the eigenvectors $\eta^{(l)}$ of $\mathscr{V}$:

$$\mathscr{V}\eta^{(l)} = \lambda_l \eta^{(l)}, \tag{4.17}$$

where, for each value of l, λ_l is a constant. Now $\mathscr{V}$ is a cyclic matrix and it seems natural to make the hypothesis that the components $\eta_m^{(l)}$ of $\eta^{(l)}$ obey the relations

$$\eta_{m+1}^{(l)} = e^{i\chi_l} \eta_m^{(l)}$$

where χ_l is a phase angle. From (4.16)

$$\eta_m^{(l)} = e^{im\chi_l} \eta_0^{(l)}. \tag{4.18}$$

Putting (4.17) in (4.15),

$$\begin{pmatrix} 2 & -1 & 0 & 0 & \dots & 0 & 0 & -1 \\ -1 & 2 & -1 & 0 & \dots & 0 & 0 & 0 \\ 0 & -1 & 2 & -1 & \dots & 0 & 0 & 0 \\ \cdot & \cdot & \cdot & \cdot & \dots & \cdot & \cdot & \cdot \\ -1 & 0 & 0 & 0 & \dots & 0 & -1 & 2 \end{pmatrix} \begin{pmatrix} 1 \\ e^{i\chi_l} \\ e^{2i\chi_l} \\ \vdots \\ e^{i(N-1)\chi_l} \end{pmatrix}$$

$$= \lambda_l \begin{pmatrix} 1 \\ e^{i\chi_l} \\ e^{2i\chi_l} \\ \cdot \\ \cdot \\ \cdot \\ e^{i(N-1)\chi_l} \end{pmatrix}$$

It is easy to verify that the system of algebraic equations (4.18) has a solution, only if

$$e^{iN\chi_l} = 1. \tag{4.19}$$

If (4.19) is satisfied, the eigenvalues λ_l will be given by

$$\lambda_l = \mathscr{K}(1-\cos\chi_l). \tag{4.20}$$

From (4.19)

$$\chi_l = 2\pi l/N, \qquad (l = 1, 2, \dots, N).$$

Hence

$$\eta_m^{(l)} = e^{2\pi i l m/N}\, \eta_0^{(l)}, \qquad (l = 1, 2, \dots, N). \tag{4.21}$$

The elements of the matrix Ω which diagonalizes $\mathscr{V}$ are precisely given by (4.21). Choosing a proper normalization constant, these elements are then

$$\Omega_m^l = \frac{1}{\sqrt{N}} e^{im\chi_l}.$$

We now introduce new coordinates q_i' related to the coordinates q_j by the relation

$$q_j = \frac{1}{\sqrt{N}} \sum_n e^{in\chi_j} q_n'. \tag{4.22}$$

Denoting by Ω^+ the matrix adjoint to Ω and using the property of Ω that it diagonalizes $\mathscr{V}$,

$$\sum_{i,j} (\Omega^+)^n_i \mathscr{V}^i_j \Omega^j_m = \lambda_n \delta_{mn},$$

where

$$\delta_{mn} = \begin{cases} 1 & \text{if } m = n \\ 0 & \text{if } m \neq n, \end{cases}$$

one finds for the potential energy (4.15),

$$\begin{aligned} V &= \sum_{i,j} q^*_i \mathscr{V}^i_j q_j \\ &= \sum_{i,j,m,n} (\Omega^+)^n_i q'^*_m \mathscr{V}^i_j \Omega^j_n q'_n \\ &= \sum_{m,n} q'^*_m q'_n \lambda_n \delta_{mn} = \sum_m q'^*_m q'_m \lambda_m \end{aligned} \tag{4.23}$$

which is now in diagonal form.

Written in terms of the new coordinates q'_m, the Lagrangian (4.13) becomes

$$L = \tfrac{1}{2}M \sum_m \dot{q}'_m \dot{q}'^*_m - \sum_m q'_m q'^*_m \lambda_m \tag{4.24}$$

and

$$p'_m = \frac{\partial L}{\partial \dot{q}'_m} = M\dot{q}'^*_m. \tag{4.25}$$

The Hamiltonian is transformed to

$$H = \tfrac{1}{2}M \sum_m p'_m p'^*_m + \sum_m q'_m q'^*_m \lambda_m. \tag{4.26}$$

The relation (4.22) can be easily inverted since Ω is unitary:

$$\sum_i (\Omega^+)^n_i \Omega^i_m = \delta_{mn}.$$

Hence

$$q'_n = \frac{1}{\sqrt{N}} \sum_j e^{-in\chi_j} q_j.$$

Since the coordinate displacements q_j are real, this shows that

$$q'^*_n = q'_{-n},$$

whence, from (4.25),

$$p'^*_n = p'_{-n}.$$

The relation (4.22) allows us to define wave numbers

$$k = \begin{cases} \chi_l/d & (0 < l \leqslant \frac{1}{2}N) \\ (2\pi - \chi_l)/d & (\frac{1}{2}N < l \leqslant N) \end{cases}$$

where d is the equilibrium distance between the point masses. In terms of these wave numbers the frequencies (4.20) become

$$\lambda_l = \mathscr{K}(1 - \cos kd).$$

We will now quantize the system. This can be done quite simply by taking p_i and q_i to be operators which obey the commutation relations

$$[q_i, p_i] = i\hbar\delta_{ij}.$$

The transformation (4.22) to the new coordinates q_i' is a canonical transformation in the sense that the q_i' and p_i' obey the same commutation relations as the q_i and p_i

$$[q_i', p_i'] = i\hbar\delta_{ij}.$$

These relations can be verified by direct substitution of (4.22) in (4.24).

The Hamiltonian (4.26) looks like the Hamiltonian of a sum of harmonic oscillators (*cf.* 4.1). It can be brought to a form similar to (4.5), i.e. to the form

$$H = \sum_k E_k(a_k^+ a_k + \tfrac{1}{2}),$$

where the operators a_k and a_k^+ obey the commutation relation

$$[a_k, a_{k'}^+] = \delta_{kk'}.$$

This, in fact, can be done immediately since the creation and annihilation operators for independent modes commute. Comparing (4.26) with (4.1) and (4.3) we find

$$H = \sum_k \hbar\omega_k(a_k^+ a_k + \tfrac{1}{2}),$$

with

$$q_k' = (a_k + a_k^+)\sqrt{\frac{\hbar}{2M\omega_k}},$$

$$p_k' = \frac{(a_k - a_k^+)}{i}\sqrt{\frac{\hbar M\omega_k}{2}},$$

$$\omega_k = \sqrt{\frac{2\lambda_k}{M}}. \qquad (4.27)$$

In the limit of long wavelengths ($kd \ll 1$), the frequency ω_k corresponding to the energy of the phonon (4.27) will be given by

$$\omega_k = \frac{E_k}{\hbar} = \left[\frac{2\mathcal{K}(1-\cos kd)}{M}\right]^{\frac{1}{2}}$$

$$\cong \left(\frac{\mathcal{K}d^2}{M}\right)^{\frac{1}{2}} k \equiv vk$$

where $$v = (\mathcal{K}d^2/M)^{\frac{1}{2}},$$

which is simply the speed of propagation of the acoustic waves in the solid, depends upon constants which characterize the solid.

The relation (4.27) is the dispersion law that we sought. In form, it is similar to the dispersion law for photons in a vacuum. The only difference is that in (4.27) it is the speed of the acoustic wave which appears, instead of the speed of light. This speed can vary from one solid to another.

4.4 Debye's Theory of the Specific Heat of a Solid

The preceding results, in particular the dispersion law, can be used to calculate the specific heat of a solid if one makes, along with Debye, two hypotheses.

The first hypothesis is that in a solid there are, for each wave vector $\vec{k}$, three modes of vibration: one longitudinal mode and two transverse modes. The speeds of propagation in the longitudinal and transverse directions are assumed to be independent of the direction and magnitude of the wave vector and will be denoted by v_L and v_T, respectively.

The second hypothesis of Debye is that the number of modes of vibration in a solid is equal to $3N$. This hypothesis is quite reasonable since this is precisely the number of vibrational modes of a system which has $3N$ degrees of freedom.

With these hypotheses, one can find an expression for the frequency spectrum of acoustic waves which, as it will turn out, does not depend upon the elastic constants of the solid.

The number of characteristic vibrations with wave vectors that have an absolute value between k and $k+dk$ can be found from the results of section 1.9 together with the well-known relation between momentum and wave vector

$$\vec{p} = \frac{h}{2\pi}\vec{k}$$

This number is

$$V\frac{4\pi k^2\,dk}{(2\pi)^3}$$

Using our first hypothesis, we find for the number of characteristic vibrations of a given type (longitudinal or transverse) in a given direction, and with frequencies between ω and $\omega+d\omega$

$$\frac{V}{(2\pi)^2}\frac{\omega^2}{v^3}\,d\omega$$

where v is the speed of this type of mode. Since there are two transverse and one longitudinal types of vibration, the total number of modes of vibration in the frequency interval between ω and $\omega+d\omega$ will be

$$\Gamma(\omega)\,d\omega = \frac{V\omega^2}{2\pi^2}\left(\frac{1}{v_L^3}+\frac{2}{v_T^3}\right)d\omega \tag{4.28}$$

Like photons, any number of phonons can occupy a given mode. Furthermore, the number of phonons is not conserved. It follows that phonons must obey Bose–Einstein statistics and that the chemical potential of a phonon gas must vanish (*cf*. section 2.12):

$$\mu = 0.$$

Hence the distribution law for phonons will be similar to the distribution law for photons

$$\bar{n}_\omega = \frac{1}{e^{\hbar\omega/kT}-1}. \tag{4.29}$$

According to Debye's second hypothesis, the maximum frequency of phonons in a crystal, $\omega_{\max}$, will be given by

$$\int_0^{\omega_{\max}}\Gamma(\omega)\,d\omega = 3N.$$

Using (4.28), we find immediately that

$$\omega_{\max}^3 = 18\pi^2\left(\frac{N}{V}\right)\left(\frac{1}{v_L^3}+\frac{2}{v_T^3}\right)^{-1}.$$

The internal energy of the phonon gas is

$$E = \int_0^{\omega_{\max}} \hbar\omega \bar{n}_\omega \, \Gamma(\omega) \, d\omega$$

$$= \frac{\hbar V}{2\pi^2}\left[\frac{1}{v_L^3}+\frac{2}{v_T^3}\right]\int_0^{\omega_{\max}} \frac{\omega^3 \, d\omega}{e^{\hbar\omega/kT}-1}.$$

Put

$$\chi \equiv \hbar\omega/kT, \qquad \Theta \equiv \hbar\omega_{\max}/kT. \tag{4.30}$$

The quantity Θ is called the *Debye characteristic temperature.* One has

$$E = \frac{\hbar V}{2\pi^2}\left[\frac{1}{v_L^3}+\frac{2}{v_T^3}\right]\left[\frac{kT}{\hbar}\right]^4 \int_0^{\chi_{\max}} \frac{\chi^3 \, d\chi}{e^\chi - 1}$$

$$= \frac{gNkT^4}{\Theta^3}\int_0^{\chi_{\max}} \frac{\chi^3 \, d\chi}{e^\chi - 1}. \tag{4.31}$$

When $T \gg \Theta$, the quantity $\chi_{\max} = \Theta/\mathrm{T}$ is small, so that we may use the approximation

$$\frac{\chi^3}{e^\chi - 1} \cong \chi^2$$

in the integral, obtaining

$$E \cong \frac{gNkT^4}{\Theta^3}\frac{\chi_{\max}^3}{3} = 3NkT,$$

which leads to the law of Dulong–Petit

$$c_V = 3Nk.$$

When $T \ll \Theta$, then $\chi_{\max} \to \infty$ and

$$\int_0^{\chi_{\max}} \frac{\chi^3 \, d\chi}{e^\chi - 1} \cong \int_0^{\infty} \frac{\chi^3 \, d\chi}{e^\chi - 1} = \frac{\pi^4}{15}.$$

Hence

$$E \cong \frac{3\pi^4}{5}\frac{Nk}{\Theta^3}T^4, \tag{4.32}$$

and the specific heat

$$c_V \cong \frac{12\pi^4}{5} Nk\left(\frac{T}{\Theta}\right)^3 \tag{4.34}$$

varies as T^3 at low temperatures. The temperature dependence of the specific heat of a solid in the low temperature region is therefore

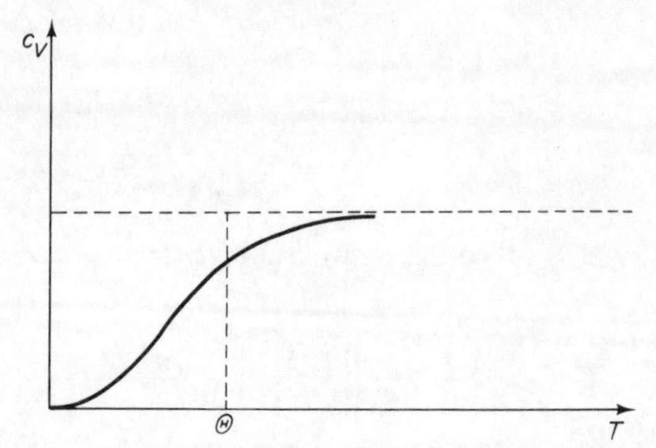

Fig. 5 The variation of the specific heat of a solid as a function of the temperature

the same as that of a photon gas. The entropy vanishes at $T = 0$ in accordance with Nernst's theorem.

The general behaviour of the specific heat of a solid as a function of the temperature, can be calculated from the general expression (4.31). This behaviour is indicated in Fig. 5.

The Debye characteristic temperature Θ can be determined by measuring the speed of sound in solids. Some experimental values are given below.

Substance	Θ (degrees Kelvin)
Be	1160
Pb	94·5
Al	418

The table shows that beryllium already behaves as a quantal solid at normal temperatures. Since the average value of Θ for most solids is of the order of 200°K, it follows that at normal temperatures and for most solids, the Dulong–Petit law is a reasonably good approximation.

Discussion The reader will have noticed that in the foregoing calculations, the symmetry properties of the atoms of the crystal were not taken into account. Only the symmetry properties of the phonons were considered. The reason can be understood as follows.

Let $\xi_1, \xi_2, \ldots, \xi_N$ be the displacements of the crystal atoms from their equilibrium positions. If $\psi_i(\xi_1, \xi_2, \ldots, \xi_N)$ denotes the normalized wave function of a state of the crystal of energy E_i, one has

$$H\psi_i = E_i\psi_i,$$

where H is the Hamiltonian of the crystal. The partition function of the solid is then

$$q = \sum_i e^{-E_i/kT} = \sum_i \langle\psi_i(\xi_1, \xi_2, \ldots, \xi_N)|e^{-H/kT}|\psi_i(\xi_1, \xi_2, \ldots \xi_N)\rangle.$$

The wave functions ψ_i will be either symmetrical or antisymmetrical, depending upon whether the crystal atoms are bosons or fermions.

But the discussion of section 2.8 showed that if the atoms are well localized, as indeed they are in most solids, i.e. if their displacements ξ_i are small compared to the interatomic distance, then the symmetry properties of the wave functions will not be relevant, because there will be practically no regions where the various coordinates ξ_i will overlap.

The problem for a liquid is entirely different because the displacements ξ_i can be quite large. This is equally true for unusual solids as, for example, solid helium. One can then understand why it is much more difficult to develop a theory of liquids (or of solid helium) than it is to develop a theory of ordinary solids.

4.5 Behaviour of Electrons in a Metal

The Bloch Theorem We shall consider here the problem of the behaviour of electrons within a conductor, assuming that the conductor has a crystal, or periodic, structure. More precisely, we assume the structure of the metal is such that there exists a three-dimensional unit cell with vectors $\vec{a}_1$, $\vec{a}_2$, and $\vec{a}_3$ which has the property that by translating it by an amount

$$\vec{R}_1 = l_1\vec{a}_1 + l_2\vec{a}_2 + l_3\vec{a}_3,$$

where l_1, l_2, l_3 are any positive or negative integers, we superimpose it exactly on some other cell of the crystal lattice. If $\vec{r}$ is an arbitrary vector, the potential energy of the system will, on account of the periodic structure of the lattice, be invariant under the translation

$$\vec{r} \rightarrow \vec{r} + \vec{R}_1$$

for any triplet $l = (l_1, l_2, l_3)$.

From this translational invariance, one can prove an important theorem on differential equations known in mathematics as Floquet's theorem,* and in physics as Bloch's theorem. We shall prove it in the particular case of the one-dimensional Schrödinger equation.

Bloch's theorem If the potential energy is invariant under the group of transformations

$$x \to x+X, \tag{4.35}$$

then the Schrödinger equation will have a solution of the form

$$\psi_k(x) = e^{ikx}\, u_k(x),$$

where k is a constant, and $u_k(x)$ is a periodic function that has the same period as the potential energy.

Proof. Let T denote the operator which effects the translation (4.35). Obviously, the kinetic energy operator of the Schrödinger equation is invariant under (4.35). Thus since, by hypothesis, T commutes with the potential energy, it commutes with the Hamiltonian

$$[T, H] = 0.$$

We can then choose solutions of the Schrödinger equation that are simultaneous eigenfunctions of H and of T

$$T\psi = \varepsilon\psi,$$

or

$$\psi(x+X) = \varepsilon\psi(x). \tag{4.36}$$

One can prove (4.36) directly. Let $f_1(x)$ and $f_2(x)$ be linearly independent solutions of the Schrödinger equation. Then any solution $\psi(x)$ is a linear combination of $f_1(x)$ and $f_2(x)$:

$$\psi(x) = Af_1(x)+Bf_2(x) \tag{4.37}$$

where A and B are constants. On account of the periodicity of the potential, $f_1(x+X)$ and $f_2(x+X)$ are also solutions of the equation and therefore

$$\begin{aligned} f_1(x+X) &= c_1 f_1(x)+c_2 f_2(x), \\ f_2(x+X) &= d_1 f_1(x)+d_2 f_2(x), \end{aligned} \tag{4.38}$$

where c_1, c_2, d_1, d_2 are constants.

We seek solutions satisfying (4.36). Using (4.37) and (4.38) we have

* E. T. Whittaker and G. N. Watson, *A Course of Modern Analysis*, Cambridge University Press (1962).

$$\psi(x+X) = (c_1 A + d_1 B) f_1(x) + (c_2 A + d_2 B) f_2(x)$$
$$= \varepsilon A f_1(x) + \varepsilon B f_2(x),$$

or

$$(c_1 - \varepsilon)A + d_1 B = 0,$$
$$c_2 A + (d_2 - \varepsilon)B = 0.$$

This system of equations has two roots with respect to ε (which may be real or complex) obtained from the condition

$$\begin{vmatrix} c_1 - \varepsilon & d_1 \\ c_2 & d_2 - \varepsilon \end{vmatrix} = 0.$$

If we define a constant λ and a function $u(x)$ such that

$$e^{\lambda X} \equiv \varepsilon, \qquad u(x) \equiv e^{-\lambda x}\psi(x),$$

then from (4.36)

$$\psi(x+X) = u(x+X)\, e^{\lambda(x+X)} = e^{\lambda X} e^{\lambda x} u(x)$$

or

$$U(x+X) = u(x).$$

Thus there exists a solution of the Schrödinger equation which has the form

$$\psi(x) = e^{\lambda x} u(x),$$

where $u(x)$ is a periodic function with the period of the potential energy. In order that ψ be normalizable, λ must be imaginary. Setting $\lambda = ik$, the equation can be written for a particular k

$$\psi_k(x) = e^{ikx} u_k(x).$$

This proves the theorem.

In three dimensions the proof of the theorem is similar and an analogous expression holds

$$\psi_k(\vec{r}) = e^{i\vec{k}.\vec{r}} u_k(\vec{r}). \tag{4.39}$$

Effective mass of electrons in a crystal An electron in a crystal is subjected to the potentials due to the other electrons, to the crystal atoms, to the phonon field, to impurities and to the exchange potential which arises as a consequence of the exclusion principle. Let $V(r)$ denote the average of the sum of these potentials. Its exact form will not be of importance to us. We shall simply assume that, inasmuch as the potential is an average potential acting on the electrons, it will be a periodic function with a period equal to that of the crystal lattice.

The Schrödinger equation for an electron of mass m moving within the potential $V(\vec{r})$ is

$$H\psi_k = E_k \psi_k, \tag{4.40}$$

where

$$H = -\frac{\hbar^2}{2m}\nabla^2 + V(\vec{r}).$$

It is of interest to compare the solution of (4.40) with the solutions of the Schrödinger equation describing an electron bound to a free atom. If $V_0(\vec{r})$ is the binding potential of the electron to the atom, the equation is

$$H_0\, u(\vec{r}) = E_0\, u(\vec{r}), \tag{4.41}$$

where

$$H_0 = -\frac{\hbar^2}{2m}\nabla^2 + V_0(\vec{r}).$$

We shall assume that the electrons are so strongly bound to their atoms that a given electron is submitted almost exclusively to the influence of the atom near which it is located. This approximation is known as the *tight-binding approximation*.

In this approximation the wave function of an electron in the neighbourhood of the jth atom will be very nearly equal to $u(\vec{r}-\vec{r}_j)$. Thus we choose for ψ a linear combination

$$\psi_k(\vec{r}) = \sum_j \alpha_j\, u(\vec{r}-\vec{r}_j).$$

The coefficients α_j can be determined from Bloch's theorem. We have

$$\begin{aligned}\psi_k(\vec{r}+\vec{R}) &= \sum_j \alpha_j\, u(\vec{r}+\vec{R}-\vec{r}_j)\\ &\equiv e^{i\vec{k}.\vec{R}} \sum_j \alpha_j e^{-i\vec{k}.\vec{R}}\, u[\vec{r}-(\vec{r}_j-\vec{R})].\end{aligned}$$

By taking $\alpha_j = e^{i\vec{k}.\vec{r}_j}$, the sum on the RHS becomes equal to $\psi_k(\vec{r})$. Thus

$$\psi_k(\vec{r}) = \sum_j e^{i\vec{k}\,\vec{r}_j}\, u(\vec{r}-\vec{r}_j) \tag{4.42}$$

obeys Bloch's theorem, and is the approximation to ψ_k that we shall use.

The energy of an electron in the crystal is given by

$$E_k = \frac{\int \psi_k^* H \psi \, d^3\vec{r}}{\int |\psi_k|^2 \, d^3\vec{r}}.$$

We have

$$H \equiv H_0 + (H - H_0) = H_0 + (V - V_0),$$

and we will treat the term $(V - V_0)$ as a small perturbation. On account of (4.42) and (4.41) we have

$$\frac{\int \psi_k^* H_0 \psi \, d^3\vec{r}}{\int |\psi_k|^2 \, d^3\vec{r}} = E_0 .$$

Thus

$$E_k = E_0 + \frac{\sum_j [\sum_j \int e^{i\vec{k}.(\vec{r}_j - \vec{r}_i)} u^*(\vec{r} - \vec{r}_i)(V - V_0) u(\vec{r} - \vec{r}_j) \, d^3\vec{r}\,]}{\int |\psi_k|^2 \, d^3\vec{r}}. \qquad (4.43)$$

From (4.42) we see that to the extent that we neglect the overlap between the atoms we can write

$$\int |\psi_k|^2 \, d^3\vec{r} \cong N,$$

where N is the number of atoms in the crystal. Furthermore, in each of the integrals in (4.43) we can make the transformation of integration variable

$$\vec{r} \to \vec{r} + \vec{r}_j$$

and introduce the quantity $\vec{\rho}_i \equiv \vec{r}_i - \vec{r}_j$. Then all the terms in the summation over j are seen to be similar and we obtain

$$E_k = E_0 + \sum_i e^{-i\vec{k}.\vec{\rho}_i} \int u^*(\vec{r} - \vec{\rho}_i)(V - V_0) u(\vec{r}) \, d^3\vec{r}. \qquad (4.44)$$

Consider the case of a cubic lattice. For simplicity we shall assume that the wave function $u(\vec{r})$ is that of a spherically symmetrical s-state and that the only integrals in (4.44) which are not negligible are those which connect nearest-neighbour atoms. Then there remain only two distinct integrals in (4.44):

$$\alpha \equiv - \int u^*(\vec{r})(V - V_0) u(\vec{r}) \, d^3\vec{r},$$

$$\gamma \equiv - \int u^*(\vec{r} - \vec{\rho})(V - V_0) u(\vec{r}) \, d^3\vec{r},$$

and

$$E_k = E_0 - \alpha - \gamma \sum_i e^{-i\vec{k}.\vec{\rho}_i}.$$

In the particular case of a simple cubic lattice, the only non-vanishing components of $\vec{\rho}_i$ will be

$$\vec{\rho}_i = (\pm d, 0, 0);\ (0, \pm d, 0);\ (0, 0, \pm d),$$

where d is the interatomic spacing. Hence

$$E_k = E_0 - \alpha - 2\gamma(\cos k_x d + \cos k_y d + \cos k_z d). \tag{4.45}$$

The above expression shows that to an energy E_0 of an electron bound to a free atom, there corresponds an energy band for the electron in the crystal

$$E_0 - \alpha - 6\gamma \leqslant E_k \leqslant E_0 - \alpha + 6\gamma.$$

The band width will be larger the greater the value of γ, i.e. the greater the overlap of the functions $u(\vec{r})$. However, one should remember that the theory applies only to strongly bound electrons, or, in other words, to the inner electrons of atoms, and not to the weakly bound valence electrons.

Suppose now that $k \ll 1$; then one can expand the cosine in (4.45)

$$E_k \cong E_0 - \alpha - 6\gamma + \gamma k^2 d^2, \qquad (k^2 = k_x^2 + k_y^2 + k_z^2). \tag{4.46}$$

The last term has the form of a kinetic energy

$$p^2/2m^* = \hbar^2 k^2/2m^* \tag{4.47}$$

if one defines an *effective mass*

$$m^* = \hbar^2/2d^2\gamma.$$

Thus we see that electrons inside a crystal that are under the influence of a periodic potential behave, for small k, as if they were free electrons with a mass equal to a certain effective mass m^*.

If $\gamma < 0$, the effective mass will be negative and when the electrons are subjected to an external electric field, they will behave as if they had a positive mass but an opposite electrical charge; i.e. they will behave as positrons. One says that the electrons behave as holes in a filled energy band. This phenomenon is often called 'hole conduction'. The electrons and the holes sometimes recombine, producing a radiative emission.

The entire theory of the fermion gas can be applied to electrons in a metal. One needs only to replace the mass of the electron by its effective mass m^*. Thus, if the parabolic approximation (4.46) is valid right up to the Fermi surface, and if the energy at the bottom of the band is taken as zero in defining μ_F, the specific heat of a degenerate gas of electrons within a metal will be given by (see equation 2.58)

$$c_V = \frac{\pi^2}{2} N \frac{k^2}{\mu_F} T, \tag{4.48}$$

and at $T = 0$, the completely degenerate fermion gas theory gives

$$N = 8\pi P_F^3 / 3\hbar^3, \tag{4.49}$$

where

$$P_F = \sqrt{(2m^* \mu_F)}.$$

The relations (4.48) and (4.49) show that a measurement of the specific heat of electrons yields the value of their effective mass. Since at very low temperatures the specific heat varies linearly with T, it cannot be mistaken for the specific heat due to the phonons, which varies as T^3.

In general, in an anisotropic solid, the kinetic energy term would have the tensor form

$$\beta_{\alpha\beta} k_\alpha k_\beta$$

instead of the simpler form (4.47). We would then obtain an effective mass tensor, instead of a scalar effective mass.

Thermionic Emission Those electrons within a conductor which have a sufficiently large kinetic energy can escape from the conductor. This phenomenon is called *thermionic emission.* The effective mass theory presented in the last subsection allows one to obtain in a straightforward fashion an expression for the current of electrons leaving the metal.

The Fermi temperature of an electron gas is (see section 2.10)

$$T_F = \frac{1}{2m^* k} \left(\frac{3h^2 \rho}{8\pi} \right)^{\frac{2}{3}}.$$

When m^* is nearly equal to the electron mass and when the density of electrons has a normal value, the Fermi temperature is extremely high. For example, if one supposes that the inter-atomic distance is of the order of one Ängstrom and that each atom has only one conduction electron, then when m^* is equal to the mass of the electron,

$$T_F/T \sim 10$$

at room temperature. Consequently, at normal temperatures, an electron gas within a metal is highly degenerate.

Let us introduce the work function

$$W = \mu_0 - \mu_F,$$

where μ_0 is the binding energy, i.e. it is the energy that is required to remove to an infinitely remote distance a zero-energy electron from the metal. The quantity W represents the energy necessary to remove from a metal an electron that has an energy equal to the Fermi energy.

When two metals are placed in contact one with the other, a current will flow between then until their Fermi levels become equal. Equilibrium will then be established. If metals A and B are, respectively, at electrical potentials V_A and V_B, then the transfer of an electron from A to B will give rise to a potential energy difference equal to $e(V_B - V_A)$ which must be equal to the difference between the work functions of A and B. Since at equilibrium $\mu_{FA} = \mu_{FB}$, we will have

$$e(V_B - V_A) = \mu_{0B} - \mu_{0A}.$$

Thus a potential difference called a 'contact potential' exists between two metals placed in contact.

Let us take the normal to the surface of the metal to be in the direction of the x-axis. Let p_x be the momentum of an electron in that direction. Then, in order for an electron to be able to leave the metal, its momentum must be such that

$$p_x^2/2m^* > \mu_0.$$

If J is the electron current per square centimeter in the direction of the x-axis, and if v_x is its velocity in that direction, one has

$$J = \frac{2e}{h^2}\int_{-\infty}^{\infty} dp_y \int_{-\infty}^{\infty} dp_z \int_{\sqrt{(2m^*\mu_0)}}^{\infty}$$

$$\times\, dp_x\, v_x \Bigg/ \left[\exp\left(\frac{p_x^2+p_y^2+p_z^2}{2m^*kT} - \frac{\mu_F}{kT}\right)+1\right]$$

$$= \frac{2e}{m^*h^3}\int_{-\infty}^{\infty} dp_y \int_{-\infty}^{\infty} dp_z \int_{\sqrt{(2m^*\mu_0)}}^{\infty}$$

$$\times dp_x\, p_x \Big/ \left[\exp\left(\frac{p_x^2 + p_y^2 + p_z^2}{2m^*kT} - \frac{\mu_F}{kT} \right) + 1 \right]$$

$$= \frac{2ekT}{h^3} \int_{-\infty}^{\infty} dp_y \int_{-\infty}^{\infty} dp_z$$

$$\times \ln\left[\exp\left(-\frac{W}{kT} \right) \exp\left(-\frac{p_y^2 + p_z^2}{2m^*kT} \right) + 1 \right]$$

The factor 2 is due to the spin degeneracy of the electrons.

At normal temperatures the first term in the logarithm is small compared to 1, which allows one to expand the logarithm and write

$$J \simeq \frac{2ekT}{h^3} \exp\left[-\frac{W}{kT} \right] \int_{-\infty}^{\infty} dp_y \int_{-\infty}^{\infty} dp_z \exp\left[-\frac{(p_y^2 + p_z^2)}{2m^*kT} \right]$$

$$= \frac{4\pi e m^* k^2 T^2}{h^3} \exp\left[-\frac{W}{kT} \right]$$

This equation is known as the Richardson–Dushman equation. One may rewrite it as follows:

$$J = c \left[\frac{m^*}{m} \right] T^2 \exp\left[-\frac{W}{kT} \right] \tag{4.51}$$

where

$$c = 4\pi e m k^2 / h^3 = 120 \text{ A cm}^{-2} \text{ deg}^{-2}$$

In general, W will be of the order of a few electron-volts.

4.6 Theory of Electrical Conductivity

Suppose that the electrons within a metal are subjected to an external electric field. On the one hand, the electrons will be accelerated in the direction of the electric field, and on the other hand they will be decelerated because they will undergo collisions either with other electrons in the metal, or with the phonons, or with the various impurities contained within the metal.

A perfectly regular and pure crystal has a high conductivity at low temperatures. The presence of impurities within the crystal, i.e. of irregularities, reduces the low-temperature conductivity in a real metal below the theoretical value for a perfect metallic crystal.

Some time after the electric field has been switched on, a new equilibrium situation will be established. To find the new distribution function, we proceed as follows.

Let $f(\vec{r}, \dot{\vec{r}}, t)$ be the number of molecules contained within a volume element $d\Omega$ in the six-dimensional space with coordinate components $r_i (i = 1, 2, 3)$ and velocity components $\dot{r}_i (i = 1, 2, 3)$. For simplicity we shall use a single label for the six coordinates

$$\chi_i = r_i \qquad (i = 1, 2, 3)$$

$$\chi_i = \dot{r}_i \qquad (i = 4, 5, 6)$$

The change of f within $d\Omega$ will come about as a result of the changes of the six coordinates χ_i due to the free motion of the particles and of the collisions between those particles which, at a given time, are contained within this volume element. We shall have

$$\frac{d}{dt}\int f(x_i)\, d\Omega = -\sum_{i=1}^{6}\int f(x_i)\dot{\chi}_i\, d\sigma_i + \int\left(\frac{\partial f}{\partial t}\right)_{\text{coll}} d\Omega \tag{4.52}$$

where $d\sigma_i$ is a surface element with a normal in the direction of χ_i, and $(df/dt)_{\text{coll}}$ is the rate of change of f due to the collisions. The region of integration is arbitrary.

The surface term above can be transformed using Gauss' theorem

$$f(\chi_i)\dot{\chi}_i\, d\sigma_i = \int \frac{\partial}{\partial \chi_i}[f(\chi_i)\dot{\chi}_i]\, d\Omega \tag{4.53}$$

Hence, combining (4.52) and (4.53) we get

$$\frac{df}{dt} = -\sum_{i=1}^{6}\frac{\partial}{\partial \chi_i}[f(\chi_i)\dot{\chi}_i] + \left(\frac{\partial f}{\partial t}\right)_{\text{coll}} \tag{4.54}$$

The above equation is knwon as the *Boltzmann transport equation*. Since we shall be interested only in its equilibrium solutions, we set

$$df/dt = 0$$

Then, expressing the RHS of (4.54) in terms of coordinates and momenta, we find

$$\vec{v}.\vec{\nabla} f + (\vec{F}.\vec{\nabla}_p)f = (\partial f/\partial t)_{\text{coll}} \tag{4.55}$$

where $\vec{F}$ is the force exerted on the electron and $\vec{\nabla}_p f$ is the gradient of f with respect to the components p_x, p_y, p_z of the momentum.

It remains to evaluate the collision term on the RHS of (4.55). Let $T_{pp'}$ be the probability that an electron makes a transition from a state of momentum $\vec{p}$ to a state of momentum $\vec{p}'$. Then from quantum mechanics one has the following expression for $T_{pp'}$:

$$T_{pp'} = \frac{2\pi}{\hbar} |H'_{pp'}|^2 \frac{V\,d^3p'}{h^3} \tag{4.56}$$

where $H'_{pp'}$ is the interaction part of the Hamiltonian that is responsible for the transition $p \to p'$. (See equation A.14 in the appendix). The particular form of $H'_{pp'}$ will depend upon the nature of the collisions, i.e. whether the collisions are between electrons, between electrons and phonons, or between electrons and impurities. Here we shall assume that there is only one type of collision and we shall leave that type unspecified.

Because we are dealing with many fermions, the formula (4.56) must be corrected by a factor which takes into account the Pauli exclusion principle. Since $f(\vec{p})$ is the probability that a state of momentum $\vec{p}$ is occupied, (4.56) must be replaced by the expression

$$T_{pp'} = \frac{2\pi}{\hbar} |H_{pp'}|^2 f(\vec{p})[1 - f(\vec{p}')] \frac{V\,d^3p'}{h^3} \tag{4.57}$$

If the state of momentum $\vec{p}'$ is occupied with probability 1, then $T_{pp'} = 0$ in agreement with the exclusion principle. The other factor $f(\vec{p})$ simply gives the initial distribution of electrons of momentum p.

Taking into account the inverse collisions, that is, those that bring electrons from a state of momentum $\vec{p}'$ to a state of momentum $\vec{p}$, and remembering that the matrix element of the Hamiltonian is symmetric

$$H'_{pp'} = H'_{p'p}$$

one has

$$\left(\frac{\partial f}{\partial t}\right)_{\text{coll}} = \frac{2\pi}{\hbar} \int |H_{pp'}|^2 \{f(\vec{p}')[1 - f(\vec{p})]\}$$

$$- f(\vec{p})[1 - f(\vec{p}')] \frac{V\,d^3p'}{h^3}$$

$$= \frac{2\pi}{\hbar} \int |H'_{pp'}|^2 \{f(\vec{p}') - f(\vec{p})\} \frac{V\,d^3p'}{h^3}$$

$$= -f(\vec{p}) \frac{2\pi}{\hbar} \int |H'_{pp'}|^2 \left\{1 - \frac{f(\vec{p}')}{f(\vec{p})}\right\} \frac{V\,d^3p'}{h^3}.$$

H

Introducing the *relaxation time* τ by the definition

$$\frac{1}{\tau(\vec{p})} \equiv \frac{2\pi}{\hbar}\int |H'_{pp'}|^2 \left\{1-\frac{f(\vec{p}\,')}{f(\vec{p})}\right\}\frac{V\,d^3p'}{h^3},$$

one finds

$$\left(\frac{\partial f(\vec{p})}{\partial t}\right)_{\text{coll}} = -\frac{f(\vec{p})}{\tau(\vec{p})}. \tag{4.58}$$

Equations (4.55) and (4.58) combine to give

$$\vec{v}.\vec{\nabla}f+(\vec{F}.\vec{\nabla}_p)f = -\frac{f(\vec{p})}{\tau(\vec{p})}.$$

Taking the electric field E to be in the direction of the z-axis, this becomes

$$\vec{v}.\vec{\nabla}f+eE\frac{\partial f}{\partial p_z} = -\frac{f(\vec{p})}{\tau(\vec{p})}.$$

In order to solve this equation we shall use the method of successive approximations. For the approximation of order zero, we take the Fermi–Dirac distribution

$$f_0 = \frac{1}{\exp[(\varepsilon-\mu)/kT]+1}.$$

The first order approximation to the distribution function f_1 will then be a solution of the equation

$$\vec{v}\,.\,\vec{\nabla}f_0+eE\frac{\partial f_0}{\partial p_z} = -\frac{f_1(\vec{p})}{\tau(\vec{p})}.$$

We shall further suppose that $\tau(\vec{p})$ depends only on the absolute value of $\vec{p}$:

$$\tau(\vec{p}) = \tau(|\vec{p}|),$$

and that the temperature T and the chemical potential μ are independent of coordinates. Since

$$\varepsilon = p^2/2m^*,$$

where m^* is the effective mass of the electron in the metal, one has

$$eE\frac{\partial f_0}{\partial p_z} = -\frac{f_1(\vec{p})}{\tau(|\vec{p}|)} \tag{4.59}$$

$$\frac{eE}{m^*}\frac{\partial f_0}{\partial \varepsilon}p_z = -\frac{f_1(\vec{p})}{\tau(|\vec{p}|)}$$

It follows that $f_1(\vec{p})$ can be written as

$$f_1(\vec{p}) = p_z\, g(\varepsilon)$$

where $g(\varepsilon)$ depends only on the energy and is given by

$$g(\varepsilon) = \frac{eE\tau(|\vec{p}|)}{m^*}\frac{\partial f_0}{\partial \varepsilon}$$

$$= \frac{eE\tau(|\vec{p}|)}{m^*kT}\,\frac{\exp[(\varepsilon-\mu)/kT]}{\{\exp[(\varepsilon-\mu)/kT+1]\}^2} \tag{4.60}$$

Thus in a first approximation, we have (using (4.59)

$$f = f_0 + f_1 = f_0 - eE\tau\, \partial f_0/\partial p_z$$
$$\cong f_0(p_x, p_y, p_z - eE\tau)$$

The electron current will be given by the following expression

$$J = e\int_{-\infty}^{\infty}\int_{-\infty}^{\infty}\int_{-\infty}^{\infty} f_0(p_x, p_y, p_z - eE\tau)\frac{p_z}{m^*}\frac{V}{h^3}\,dp_x\,dp_y\,dp_z$$

$$= \frac{e}{m^*}\int_{-\infty}^{\infty}\int_{-\infty}^{\infty}\int_{-\infty}^{\infty} p_z\left[\exp\left(\frac{p_x^2+p_y^2+(p_z-eE\tau)^2}{2m^*kT}-\frac{\mu}{kT}\right)+1\right]^{-1}\frac{V}{h^3}$$
$$\times\, dp_x\,dp_y\,dp_z\,.$$

By making the transformation of variables

$$p_z' = p_z - eE\tau$$

the above expression becomes

$$\mathrm{J} = \frac{e}{m^*}\int_{-\infty}^{\infty}\int_{-\infty}^{\infty}\int_{-\infty}^{\infty} (p_z' + eE\tau)\left[\exp\left(\frac{p_x^2+p_y^2+p_z'^2}{2m^*kT}-\frac{\mu}{kT}\right)+1\right]^{-1}\frac{V}{h^3}$$
$$\times\, dp_x\,dp_y\,dp_z\,.$$

The term in the integrand that contains p_z' is an odd function of p_z' and gives no contribution to the integral; there remains

$$J = \frac{e^2 E\tau}{m^*} \int_{-\infty}^{\infty} \int_{-\infty}^{\infty} \int_{-\infty}^{\infty} \left[\exp\left(\frac{p_x^2 + p_y^2 + p_z'^2}{\lambda} - \frac{\mu}{kT} \right) + 1 \right]^{-1} \frac{V}{h^3}$$
$$\times \, dp_x \, dp_y \, dp'_z .$$

The integral above is nothing else but the average number of particles. Thus finally

$$J = e^2 EN\tau/m^* \tag{4.62}$$

and the electrical conductivity σ is

$$\sigma = J/E = e^2 N\tau/m^* \tag{4.63}$$

This expression is similar in form to the classical Drude–Lorentz equation $e^2 N\tau/m$. One must remember, however, that the ratio τ/m^* should be calculated using quantum mechanics.

A simplifying feature appears at very low temperatures. Since τ is proportional to the function $g(\varepsilon)$ (4.60), which becomes extremely peaked around the Fermi energy as $T \to 0$, we need only to know the value of the relaxation time for $E = \mu_F$ at temperatures near zero.

EXERCISES

1 Calculate the number of phonons in a solid as a function of temperature.

2 Find the first order solution of the Boltzmann equation when both the temperature and the chemical potential are coordinate dependent.

3 The thermal conductivity of a solid at a temperature T is defined as

$$\kappa = Q_i/|\Delta T|$$

where Q_i is the thermal current in the i direction

$$Q_i = \int v_i \varepsilon f \, d^3 v$$

and ΔT is the temperature gradient in the same direction. (v_i is the velocity of electrons in the i direction, ε their energy, and f the equilibrium distribution function).

(a) Determine the electric field E_z in the z-direction which makes the electric current J_z vanish.

(b) Using the value of E_z found above, determine κ for low T, and show that the ratio of thermal to electrical conductivity obeys the Wiedemann–Franz law

$$\frac{\kappa}{\sigma} = \frac{\pi^2}{3}\left(\frac{k}{e}\right)^2 T$$

where e is the electric charge and k the Boltzmann constant.

4 Consider a solid spontaneously emitting photons. Find the temperature dependence of the thermal conductivity due to the photons when

$$pc/kT \ll 1$$

where p is the photon momentum.

APPENDIX

We shall present here the derivation of the master equation as it was originally given by Pauli [14]. The hypotheses introduced by Pauli are stronger than those that are actually needed, as shown by more recent derivations. However, Pauli's demonstration has the double merit of being simple and of illustrating quite adequately the problems that enter.

The Schrödinger equation for a system of isolated particles is

$$(H_0+V)\psi(\vec{r},t) = i\hbar\,\partial\psi(r,t)/\partial t \tag{A.1}$$

where H_0 is the unperturbed Hamiltonian and V is that part of the Hamiltonian that gives rise to interactions between the particles. Let $\psi_m(\vec{r})$ be an eigenfunction of H_0 corresponding to the energy E_m:

$$H_0\,\psi_m(\vec{r}) = E_m\,\psi_m(\vec{r}) \tag{A.2}$$

The solutions of (A.1) in the absence of an interaction are of the form

$$\psi_m \exp(-i/\hbar E_m t) \tag{A.3}$$

and the general solution of (A.1) for $V \neq 0$ can be written as a linear combination of solutions of the type (A.3),

$$\psi(\vec{r},t) = \sum_m c_m(t)\psi_m(\vec{r})\exp(-i/\hbar E_m t) \tag{A.4}$$

Putting (A.4) into (A.1), multiplying the resulting equation by $\psi m^*(\vec{r})$ and integrating over $\vec{r}$, one obtains a system of differential equations for the coefficients $c_m(t)$

$$\frac{\hbar}{i}\frac{dc_m(t)}{dt} = \sum_m V_{mn}\exp\{i/\hbar(E_n-E_m)t\}c_n(t) \tag{A.5}$$

where

$$V_{mn} \equiv \int \psi m^*(\vec{r})V\psi_n(\vec{r})\,d^3r \tag{A.6}$$

obeys the symmetry relation

$$V_{mn} = V^*_{nm} \tag{A.7}$$

since the potential V is hermitian.

The system of differential equations (A.5) is equivalent to the

Schrödinger equation. If now we assume that the potential V acts as a small perturbation, the matrix elements (A.6) will be small, and the coefficients $c_m(t)$ will be slowly-varying functions of the time. One can then, in a first approximation, integrate the system of equations (A.5) over a time interval that is sufficiently short so that the coefficients $c_n(t)$ on the RHS of (A.5) can be replaced by their average values $c_n(0)$ at $t \approx 0$.

The system of equations now becomes easily integrable. The solution is

$$c_m(t) = \sum_n V_{mn} \frac{[\exp\{i/\hbar(E_m - E_n)t\} - 1]}{(E_n - E_m)} c_n(0) \tag{A.8}$$

where the sum extends over all states that are present initially (we are supposing that m is a *final* state). Taking the square of the modulus of (A.8), we have

$$|c_m(t)|^2 = \sum_{n,l} V^*_{ml} V_{mn}$$

$$\frac{[\exp\{i/\hbar(E_m - E_n)t\} - 1][\exp\{-i/\hbar(E_m - E_l)t\} - 1]}{(E_n - E_m)(E_l - E_m)} c_n(0)\, c^*_l(0) \tag{A.9}$$

At this point we must make some hypotheses in order to be able to evaluate the sum above. Let us suppose that at $t = 0$, the system can be in any one among Ω groups of initial states and let P_i be the probability that the system is in one of the states belonging to group i. Following Pauli, we shall write for the average value of the product of the coefficients that appear in (A.9):

$$\overline{c_n(0)\, c^*_l(0)} = \begin{cases} \delta_{lm} P_i/\omega_i & \text{if } l \text{ belongs to group } i \\ 0 & \text{otherwise} \end{cases} \tag{A.10}$$

In order to arrive at (A.10) one must assume

(a) that the phases of the wave function for $t \simeq 0$ vary randomly, so that the non-diagonal terms vanish on average,

(b) that the occupation probabilities of each of the states that belongs to a given group of initial states of the system are, *a priori*, equal.

The first postulate is called the random phase approximation;

the second postulate is called the postulate of equal *a priori* probabilities.

With these hypotheses, we obtain for the transition probability to a group of final states j,

$$p_i' = \sum_m |c_m(t)|^2$$

$$= 2\left\{\sum_{m,n} |V_{mn}|^2 \frac{1-\cos[(E_m - E_n)t/\hbar]}{(E_m - E_n)^2}\right\}\frac{p_i}{\omega_i} \tag{A.11}$$

where m is an index which refers to all accessible states of the system that belong to the group of final states j.

If the states n and m belong to a continuous group of states with energies in the range $E \pm \Delta E$, the sums above can be replaced by integrals. Let $\rho_n(E)$ and $\rho_m(E)$ be the density of energy levels of the groups to which the states n and m belong. The most important terms in (A.11) are those which come from energy conserving transitions: $E_m \approx E_n$. If we assume that these densities are slowly varying functions of the energy in the energy range where they are defined, and that $|V_{mn}|^2$ is approximately constant in that energy range, then (A.11) becomes

$$p_j' = \frac{2P_i}{\omega_i}|V_{mn}|^2 \rho_n(E)\rho_m(E) \int_{E-\Delta E}^{E+\Delta E} \int_{E-\Delta E}^{E+\Delta E}$$

$$\times \frac{1-\cos[(E_m - E_n)t/\hbar]}{(E_m - E_n)^2} dE_m\, dE_n$$

putting

$$y \equiv (E_m - E_n)t/\hbar$$

we get

$$p_j' = \frac{2P_i|V_{mn}|^2}{\omega_i \quad \hbar} \rho_n(E)\rho_m(E) \int_{E-\Delta E}^{E+\Delta E} \int_{-(\Delta E)t/\hbar}^{+(\Delta E)t/\hbar} \frac{(1-\cos y)}{y^2} dE_n\, dy$$

For times

$$t \gg \hbar/\Delta E$$

the integral over y can be extended over the entire range $(-\infty, +\infty)$, and since

$$\int_{-\infty}^{\infty} \frac{(1-\cos y)}{y^2}\, dy = \pi,$$

we find

$$p'_j = \frac{2P_i \pi}{\omega_i \hbar} |V_{mn}|^2 \rho_n(E) \rho_m(E) (\Delta E) t \tag{A.12}$$

The probability per unit time R_{ij} that the system makes a transition from a group of states i to a group of states j is therefore

$$R_{ij} = \frac{p'_j}{t} \equiv \frac{T_{ij} P_i}{\omega_i} \tag{A.13}$$

where

$$T_{ji} = T_{ij} = \frac{2\pi}{\hbar} |V_{mn}|^2 \rho_n(E) \rho_m(E) \Delta E \qquad (m, n \in i) \tag{A.14}$$

The time variation of the occupation probability of the group of states i is obviously given by

$$dP_i/dt = \sum_{j \neq i} (R_{ji} - R_{ij})$$

Putting

$$R_{ij} \equiv \mathscr{H}_{ij} P_i$$

where

$$\mathscr{H}_{ij} = T_{ij}/\omega_i$$

has the symmetry property

$$\mathscr{H}_{ij} \omega_i = \mathscr{H}_{ji} \omega_j$$

we have

$$dP_i/dt = \sum_j (\mathscr{H}_{ji} P_j - \mathscr{H}_{ij} P_i)$$

Introducing indices r and s which refer now to individual states, rather than to groups of states, and defining the quantities

$$p_r = \frac{P_i}{\omega_i} \quad \text{and} \quad H_{rs} = \frac{\mathscr{H}_{ij}}{\omega_j}$$

we are lead to

$$dp_r/dt = \sum_s H_{rs} (p_s - p_r) \tag{A.15}$$

with $H_{rs} = H_{sr}$.
Equation (A.15) is the master equation.

In order to obtain the master equation, we had recourse to perturbation theory. This theory is valid for small t because we assumed that the probabilities were not strongly time-dependent. In order for the master equation to be valid for large values of t, one must repeatedly introduce the random phase approximation for a whole sequence of values of t, separated one from the other by sufficiently small time intervals. This is tantamount to assuming that the phase of the wave function of the system varies randomly at all times.

This hypothesis is too strong. More recently L. Van Hove has shown by a non-elementary calculation [19, 20] that it is sufficient to introduce the random phase approximation for the wave function of the system, at the initial time only. Unfortunately, it is not possible to give here a detailed account of Van Hove's derivation.

REFERENCES

1. BOER, J. DE, *Reports on Progress in Physics,* **12**, 305, 1948–49.
2. CHANDRASEKHAR, S., *Stellar Structure,* Dover Publications Inc., 1957.
3. DAVIDSON, N., *Statistical Mechanics,* McGraw-Hill, New York, 1962.
4. DENNERY, P. and KRZYWICKI, A., *Mathematics for Physicists.* Harper and Row, New York, 1967.
5. DIRAC, P. A. M., *The Principles of Quantum Mechanics,* Oxford, Clarendon Press, 4th ed., 1958.
6. GIBBS, J. W., *Elementary Principles in Statistical Mechanics,* Collected Works, New Haven, 1948.
7. HILL, T. L., *An Introduction to Statistical Thermodynamics,* Addison-Wesley, Reading Mass., 1960.
8. HUANG, K., *Statistical Mechanics,* John Wiley and Sons, New York, 1965.
9. KAHN, B. and UHLENBECK, G. E., *Physica,* **5**, 399, 1938.
10. KITTEL, C., *Elementary Statistical Physics,* John Wiley and Sons, New York, 1961.
11. KITTEL, C., *Introduction to Solid State Physics,* John Wiley and Sons, New York, 3rd ed., 1966.
12. LANDAU, L. D. and LIFSHITZ, E. M., *Statistical Physics,* Pergamon Press, London–Paris, 1958.
13. MUNSTER, A., *Handbuch der Physik,* Springer-Verlag, Berlin, 1959.
14. PAULI, W., *Collected Scientific Papers,* Vol. 1, p. 549, Interscience; New York and London, 1965.
15. SHER, A. and PRIMAKOFF, H., *Phys. Rev.,* **119**, 178, 1960.
16. TER HAAR, D., *Elements of Statistical Mechanics,* Holt, Rinehart and Winston, New York, 1961.
17. TER HAAR, D., *Rev. Mod. Phys.,* **27**, 289, 1955.
18. THOMSON, R. M., *Quantum Theory* II (*Aggregates of Particles*), Academic Press; New York and London, 1962.
19. VAN HOVE, L., *Rendiconti* S.I.F. XIV Corso, p. 155–169.
20. VAN HOVE, L., *Physica* **21**, 517, 1955; *Physica* **23**, 441, 1957.
21. VAN KAMPEN, N. G., *Physica,* **27**, 783, 1961.

HINTS AND ANSWERS TO SELECTED EXERCISES

Chapter 1

3 Show that $V(t) = V(0)$ where $V(0)$ is the same integral as $V(t)$ with, however, $p_i, q_i\,(i = 1, 2, \ldots, n)$ replaced by their initial values $P_{i_0}, q_{i_0}\,(i = 1, 2, \ldots, n)$.

4 Study the variation of the temperature using the definition

$$\frac{1}{T} = \left(\frac{\partial S}{\partial E}\right)_{V,N}$$

Chapter 2

2 The entropy is given by

$$S = k \ln \frac{\Gamma}{h^{3N}}$$

where

$$\Gamma = \frac{V^N}{N!} \int \prod_{j=1}^{3N} dp_j$$

The integral is subject to the constraint

$$0 - \Delta E \leqslant \frac{1}{2m} \sum_{j=1}^{3N} p_j^2 \leqslant E$$

which defines a shell of thickness ΔE on a $3N$-dimensional hypersphere. Show that this constraint is equivalent to the constraint

$$0 \leqslant \frac{1}{2m} \sum_{j=1}^{3N} p_j^2 \leqslant E$$

when N is very large, which in turn defines the volume of a hypersphere of radius $p = \sqrt{(2mE)}$ and of volume proportional to p^{3N}. The energy is obtained from the relation

$$\frac{1}{T} = \left(\frac{\partial S}{\partial E}\right)_{V,N}$$

3 The partition functions are given by

$$q = \frac{1}{2}\left(\frac{1}{1-e^{-\hbar\omega/kT}}\right)^2 \mp \frac{1}{2}\left(\frac{1}{1-e^{-2\hbar\omega/kT}}\right) \qquad \begin{matrix} -\text{ for fermions} \\ +\text{ for bosons} \end{matrix}$$

$$q = \left(\frac{1}{1-e^{-\hbar\omega/kT}}\right)^2 \qquad \text{for distinguishable particles}$$

4 $\overline{c^2p_i^2/E} = kT \qquad (i = x, y, z)$

6 No.

Chapter 3

1 $\frac{4}{3}L^2$

Chapter 4

2 If the temperature and the chemical potential depend on the coordinate x, then

$$f_1 = \frac{\tau v_x}{kT}\left[cE_x + \left(\frac{\varepsilon-\mu}{T} + \frac{\partial\mu}{\partial T}\right)\frac{\partial T}{\partial x}\right]\frac{\partial f_0}{\partial\alpha}$$

where $\alpha = (\varepsilon-\mu)/kT$ and τ is the relaxation time.

3 $-\partial f/\partial\varepsilon$ behaves very nearly as a δ-function for small T.

4 Independent of T.

Index